T0244067

THE LAST SUPPER CLUB

THE LAST SUPPER CLUB

A Waiter's Requiem

MATTHEW BATT

UNIVERSITY OF MINNESOTA PRESS

MINNEAPOLIS • LONDON

Published by the University of Minnesota Press
111 Third Avenue South, Suite 290
Minneapolis, MN 55401-2520
http://www.upress.umn.edu

ISBN 978-1-5179-1485-1 (hc)
ISBN 978-1-5179-1615-2 (pb)

LC record available at https://lccn.loc.gov/2023033989

Printed in the United States of America on acid-free paper

The University of Minnesota is an equal-opportunity educator and employer.

30 29 28 27 26 25 24 23   10 9 8 7 6 5 4 3 2 1

For Jenae and Emory, again and again

Venite ad me omnes qui stomacho laboretis,
et ego restaurabo vos.

Come unto me, all you whose stomachs are aching,
and I will restore you.

—M. BOULANGER, 1765

Contents

PROLOGUE The Last Supper........................ 1

1. Duck Tongues..................................... 5

2. Curing the Heart................................. 20

3. Heaven City...................................... 41

4. Schadenfreude................................... 57

5. Hunger Games.................................... 73

6. The Pregame Game Show 87

7. The German Word for Light 100

8. Imaginary Beer................................. 116

9. Welcome to the Occupation 133

10. Part-Time Pirates............................... 151

11. The Taste of Metal 166

12. Immediate Seating.............................. 182

13. Duck, Duck, Gray Duck 198

14. No Place for Pity............................... 213

15. Salt.. 232

EPILOGUE Beautiful Friends 245

ACKNOWLEDGMENTS 261

Prologue

THE LAST SUPPER

IT'S AN HOUR BEFORE OUR LAST SERVICE BEGINS, AND JORGE is leaning against the balcony rail above the beer garden below, and he's poised like a small god, reminding me suddenly of Kurtz as he surveys his strange and fragile empire just before it—and he—is consumed by the wilderness. Jorge is bald but has a grizzled black and gray beard that gives him an air of power, wisdom, and time having been served in one institution or another. Maybe prison, maybe professional cage fighting, or maybe just lots and lots of time in restaurants. Hundreds of people loiter below, some waiting in line for beer, some playing a game with beanbags called cornhole, others just eating and drinking with their friends, family, kids, and dogs. Hundreds of them and it's just the middle of the afternoon—usually any restaurant's deadest time of day—but this place is as busy as a hub airport at peak rush hour. Jorge is the executive chef of both the beer hall and garden below and our restaurant on this second floor, The Brewer's Table. He's standing off to the side of the balcony, away from the last family meal he just made for us all. It's his favorite dish, frijoles con puerco, the one he used to beg his Yucatecan grandmother to make.

"It's super simple," he said when he served us. He was choking up, but he muscled through as he rubbed the big rooster tattoo on his forearm. "But it's so delicious. I can't talk about it or I'm going to fucking cry. And the mojo rojo is super spicy. Seriously. Don't fuck around with it or it'll burn your face off. Enjoy."

We put seven tables together on the balcony so that we could, for one last time, all eat together. We didn't always have time for family

I

meal, but when we did, it invariably made the night better. Often the chefs would have only ten minutes or so before service was to begin to cook for themselves and the front-of-house staff, and they'd whip up stir-fries or fish tacos or these crazy casseroles and line them up in hotel pans or on sheet trays right on the pass for everybody who's got time to dig in and chow down, all while belting out one Whitney Houston or Nine Inch Nails song after another. And while we often ate at the same time, until tonight, at this big, long plank of a table, we have never eaten, all of us, together.

Today, everybody piles up their frijoles, grabs some warm tortillas, limes, a squirt of chipotle crema, a couple of drops of Jorge's mojo, and finds a seat at the long table. We eat and chat and try to laugh at Pleezer, the Weezer cover band playing below, because the only other option is to openly weep, and there will be time enough for that later.

"So if this is the last supper," Baby Steps says, "who's Jesus and who's Judas?"

Baby Steps is one of our slow but methodical bartenders. He doesn't drink anymore, and as he floats his question he's gifting each of us rare bottles of beer and wine that he has stockpiled over the years. He's trying to be funny with his question, I think, to undermine how sweet his gesture is, but it's a big question and a big, sweet gesture. As he comes around, he hands me an ale brewed in Belgium by Trappist monks that's sold only at their monastery. One eleven-ounce bottle is probably worth a hundred bucks—if you could even find one on this continent. They don't even bother putting labels on their bottles.

Everybody laughs at Baby Steps's joke-question, but nobody really replies. We all know Jorge is Jesus and Omar—the founder and owner of the Surly Brewing Company and this whole destination brewery and pair of restaurants—no question, Omar is Judas.

After a two-and-a-half-year run that included a Cochon 555 victory, two James Beard nominations as well as a finalist nod for Best Chef in the Midwest, and even a coveted place on *Food & Wine* magazine's very short list of the best restaurants in America, Omar is shutting us down.

The party line he gave us a few months ago at an all-staff meeting on a chilly day in May felt like a bunch of corporate-speak horseshit. "The Surly Brewing Company, you know," he said, his eyes darting once around the room and then settling on the tops of his shoes. "Well, we've always been at the forefront of innovation, and, having accomplished the goals we set out for ourselves, we've decided to challenge ourselves with something new . . ."

He left the meeting quickly without shaking anybody's hand.

Most guessed it came down to a pissing contest between Omar and Jorge. Profit margins, of course, were likely also a big part of the decision. It might just be the new corporate manager from the Granite City conglomeration who took over the beer hall recently. Nobody I ever talked to actually knew the real reason or if they did wouldn't say much about it.

All any of us really know is that today is it. The end.

Before we clean up from our meal and get ready for our last service, the whole front- and back-of-house staff gathers for one last picture together.

"All right," Emily, our manager, says, framing us all on the screen of her phone. The hum of the crowd below begins to grow as Pleezer nears the chorus of "The Sweater Song."

"On three," Emily says. "One, two, three!"

"Queso!" we all say, and every one of us gives Omar the finger.

Chapter One

DUCK TONGUES

THE INTERVIEW STARTS IN A LITTLE UNDER AN HOUR, AND I don't know what to do with myself. I've sent my resume and filled out the application. I've exchanged pleasant emails with their HR person. I've researched the company's founder, its origins, its market share— all that nonsense I really know nothing about. I'm an English major, after all, and I'm applying for a bartending job at a brewery.

But that's not quite right.

I was an English major.

Now I am an associate professor of English at Minnesota's largest private university. But I'm on sabbatical, and I'm damned near broke. I thought I could get some freelance work as a copy editor or an adjunct instructor or even try to get a slot teaching at The Loft, one of the nation's finest creative writing labs for non-degree-seeking students. But I don't want any more teaching for now—also, I'm pretty sure there was language in my contract preventing me from teaching elsewhere within academia while on sabbatical. Language preventing me from working at a job most of my students deem beneath them, however, was absent.

I've been teaching at a college for fifteen years, and I'm cooked. It's a great job, no doubt about it. Solid benefits. The summers off and the January term when I typically don't teach are worth whatever deficits come with working in higher education. And sabbatical thus far has been astonishingly humane and brilliant. I finished work on one novel and completed a draft of another. More important, I've been taking naps and running more and going to the climbing gym as though I'm still some free-range graduate student.

But I'm broke.

I don't know about all schools, but where I teach, a yearlong sabbatical comes at a 50 percent reduction in pay. I didn't think I'd feel the pinch when I signed up for it. Nine months in, it doesn't feel so much like a pinch but rather like I've been dipped in chum and thrown in a tub of piranhas. The math was obvious and easy, but like I said, I was an English major. Everything that involves numbers—reading a calendar, even trying to recall my son's birthday—brings me right back to all the math trauma I barely survived in high school and college.

So I went as long as I could on fumes, but my salary won't return to its normal pay grade for many months, and I've got to do something about it now. I went from making probably what an assistant manager at Target earns to now making less than the kid who delivers our paper. On top of all that, I've got student loans that just about rival Ireland's national debt. What most folks don't know about people who teach in higher education is that, in most cases, we've moved several times, several states away from wherever home was, chasing advanced degrees, fellowships, teaching assistant gigs, and adjunct faculty jobs—most of which tend to pay, literally, less than minimum wage—all in pursuit of the mythical golden goose of academia: a tenure-track job.

Since my undergraduate days, I've lived in Wisconsin, Massachusetts, Ohio, back to Wisconsin, on to Utah, Texas, and (finally?) my new home in Minnesota. That all costs money. A lot of money.

As a result, I have no savings account, single digits left in our checking account, nearly a hundred grand in student loans that have already been in forbearance, a car payment that will probably outlast the functionality of the vehicle, tens of thousands of dollars in credit card debt, and a monster mortgage from buying a house at the worst time in recent history.

Foreclosure and bankruptcy are damned tempting, but like most Americans, I prefer to work hard rather than ask for preferential treatment. And so long as we're willing, our debtors are more than happy to let us keep paying. And in order for me to keep paying, it's either land a monster book contract or suck it up and get a second job.

Suffice it to say, the book contract thing hasn't happened yet.

* * *

At the beginning of my sabbatical I had planned on writing a book about work in America. Driven by immersion projects like Ted Conover's and, in particular, Barbara Ehrenreich's *Nickel and Dimed*, where she discovered and dramatized how shitty jobs pay shitty wages, I had wanted to do the opposite: figure out which are the best jobs in America that you don't need advanced degrees to land but that still pay well enough to make ends meet *and* people are typically happy to do. I had done a bunch of research and found jobs both easy to guess (primary school teacher, artisan woodworker) and unexpected (heavy equipment operator, firefighter) and totally unbelievable (financial planner). I had planned to try to work some of these jobs, each for a month at a time, *Super Size Me*–style like Morgan Spurlock, but I got caught up in the novels I had been working on and never really found the right path in or the time frame to work a bunch of different jobs in a meaningful way within the few months of my sabbatical.

Then I discovered abruptly that I had to pay either my student loans or my son's day care bill or not put food on the table. So, out the window went the novels and the book about work. I needed a job and I needed one now and it had to pay quick. That meant only one thing to me: a restaurant job.

I've always loved working in restaurants. I did so the entire time I was an undergraduate and graduate student and have never lost the passion for it. As I learned how to become a writer and a teacher, I also learned more and more about how to be a better server, working in no fewer than five restaurants over the thirteen years I spent in various colleges. And during this time, when I would grow exhausted of my own writing and scholarship, I never tired of the restaurant world in all its forms. Not only did I love eating out and working as a server and sometimes a bartender and very occasionally a cook, but I would also read chef memoirs. I watched all the foodie TV shows and documentaries, even the treacly chef biopic thing starring Jon Favreau. Even *Diners, Drive-Ins, and Dives*. Especially *Kitchen Confidential* and both of Anthony Bourdain's shows. I've watched all 246 episodes of *No Reservations* and *Parts*

Unknown at least twice. The Montreal one with the Joe Beef guys probably five times. The Beirut one I can't count how many.

Nobody better bridged the gap between restaurants and the creative writing world better than Bourdain, and while I knew it was unlikely he'd ever need an apprentice, I dreamed that my career arc could be kind of the opposite of his own. Instead of working in the kitchen, I'd work the floor, and instead of starting out my creative career by going to culinary school, I'd go to an MFA program, and then we'd meet in the middle, launching what would be our life's real work by writing an honest book about working in restaurants.

But, Bourdain is now gone, and I'm not applying for a restaurant job so I can write about it. First and foremost, I need the money now. I could go on and on about how much I love it, but what I also love is the fast, honest money—you work hard, you make good tips; you work harder, you make better tips—and in a state like Minnesota where servers make minimum wage *and* tips, it's actually one of the best places in the world to be a server, so back into the fray I decided I would wend.

I'm at the Half-Time Rec, my neighborhood dive bar famous for a cameo in *Grumpy Old Men*, spending the last few single bills that I could scrounge together on a burger and a beer when I decide to start looking for my own restaurant job. It quickly becomes apparent that these days, for restaurants, Craigslist is it. There's no LinkedIn for waiters. No Monster.com for busboys. Just the free, digital equivalent of the want ads in the back of the paper.

I'm finding that only the newest or most corporate of restaurants list their names in their posts. The need to advertise for jobs in the dining industry is, I'm gathering, one of the early tells of a terminal illness within the business model of your restaurant. Thinking back on it, the higher-end/higher-quality places I've worked at always had applicants just showing up at the host stand, day after day, often right in the middle of a busy shift (those applicants were rejected on the spot; if you don't have the sense to observe when a bad time to ask about employment is, the restaurant industry has a very direct way of teaching you all about it). It's easy to tell which restaurants are the busiest. If you can't get a reservation there, the server's ideal model of employment goes, try to get a job there.

Looking over the postings as Patrick, the erstwhile Ray Liotta stunt double tending bar at the Rec, pours me what has to be my last tap beer, I notice on my phone that there are more than a few coy postings ("Hot night spot in Uptown needs qualified waitstaff. Big $$$!" "Don't bother to apply if you don't got the skil$$!"). Dollar signs and anonymity are rife in this sub-universe, but I'm willing to wager that they're more than overcompensating.

I'll work anywhere the customers are, but my only condition is that I don't want much, if any, overlap between my students and the guests. I recall that happening once at the last place I worked. An ironically well-to-do long-haul trucker student of mine at the University of Utah got wind that I worked at the New Yorker, a high-end place in downtown Salt Lake City, and he brought his girlfriend and requested me specifically. It was fine: he knew he could afford to be there, but I could tell by the way his doe-eyed girlfriend cowered against him the entire meal that she wasn't at all comfortable in our marble-and-brass-detailed dining room. It was just plain awkward, and I'm not eager for it to happen again.

So, what's out? Pretty much any place south of I-94, the interstate that divides Saint Paul more or less in two, where my university is and I'm therefore more likely to run into my students. Downtown Saint Paul is pretty much a snoozefest with a couple of notable exceptions, but I don't want to work someplace where the only nights you can bank on business are when there are twenty thousand hockey fans in town all looking for somewhere to eat and get toasted before the game. There are many places in the North Loop and the Uptown part of Minneapolis, including one right next door to the climbing gym I frequent, but I kind of want to keep that as mine. I don't know. Working there would be cool, I guess—I'd certainly get to know everybody in the area—but it's a bit too hipster ridden for my taste/age.

Then it hits me.

I look so astonished that Patrick abandons the rack of cinnamon whiskey shots he had been pouring for a bunch of bachelorette types. "You all right, pal?"

"Yeah," I say. "I'm good. Or at least I think I will be."

As Patrick shrugs and resumes his shot pouring, I think, Where have I been nearly a dozen times since it opened? Where have I been on back-to-back days? Where have I been and happily waited for over an hour for a table? Where have I complimented the bartenders and the wait-staff on their kind, attentive, and swift service? Only one place in recent memory.

The new Surly beer hall.

Just three miles from my house, it's located halfway between Saint Paul and Minneapolis, right off the bus/bike-only intercampus connector that could get me there in twenty minutes under my own power, probably about the same amount of time it would take to drive. Ever since it opened the December prior, it's been absolutely jammed every time I've been there—and I've never gone on a Friday or Saturday night, when I imagine the wait times run upward of two or three hours. I'd go there for the beer alone. I've wanted to tour Surly's Brooklyn Center brewing facility nearly an hour away from my house, but the prospect of needing a hundred-dollar cab ride home after all the free samples has kept me away.

But now the new beer hall is just a two-wheeled ramble away. It took a while before my wife and I went the first time—I don't really know why—but once we did, a seal was broken.

Not only do they have the best beer in the world, but the space is amazing. They repurposed and built on a wasteland industrial lot where nothing but postapocalyptic, disused grain elevators stubble the landscape like tombstones to the titans, and they turned the space into what can reservedly be described as a cathedral to beer. The place is like Valhalla, full of awe and majesty, but it also has a sense of humor mixed with a kind of fury. As you enter the building, you pass an open fire pit alive with gas-fed flames bigger than any witch's cauldron, and inside you're immediately struck by the marriage of dark hardwood floors and walls up against the gleaming stainless steel of the brewing fermenters, tall and as imposing as Pershing missile silos.

Down a wide and welcoming hall, with the din of the dining room now abundantly audible, you pass a dozen or so original lithographs celebrating the brewery's divine and vaguely heavy metal–inspired beers. The dark and horrifying-but-in-a-good-way Blakkr (a winey, dark, double IPA), CynicAle (a refreshing and vibrant Weiss beer), Misanthrope (a wild-yeast sour), and, my favorite, Furious (a British ESB/American IPA), to name just a few.

And then you come to what feels like the mouth of a cave, and it's impossible to not gape as you take in the enormity of the space. From floor to ceiling, it's probably forty church-like feet, and the view across the nearly acre-wide space is unimpeded by any support beams or walls, and starlike lights glitter the ceiling in huge wheel-shaped fixtures. On one end of the dining room the bright silvery white of yet more gleaming beer tanks shine, and at the other end lights illuminate a spacious outdoor beer garden enclosed by ember-red-glowing grain elevators.

One glance at their dining hall tells you that this is a finely tuned operation, and I'm betting it's also one where everybody is pretty much always dying to work if they aren't already. There's no way I'm going to find a listing for this place on Craigslist, I think, nursing a sip of my tap Summit EPA. But I check their website, and they are indeed hiring! Of course a restaurant this busy doesn't even need to post listings on Craigslist.

But the polish of the news immediately wears thin. The two positions they're advertising for are host (the person who greets you, tells you the grim news about how many hours you'll be waiting, and, if attrition doesn't get the better of you, eventually shows you to your table) and, worse yet, in my snotty estimation, banquet staff.

Other than the private events at the restaurants I've worked at, I've never done banquets, but my wife did millions back when she was an event manager. It was easily the job that stressed her out the most and the one I'm the least eager to wade into. Think about it. A crew of folks who don't know what to expect day to day put up against the insanely voracious expectations of perfection by red-eyed brides, their mothers, and/or corporate fascists.

Add that to the fact that all of the major events in anybody's given life are the days that have stolen my mother, a florist, away from me during my entire life: Christmas, Easter, Mother's Day, proms, graduations, weddings, funerals. You name it. If it's a big day you're looking forward to/dreading, it's something that meant my mom was working yet another fourteen-hour day to make your stupid boutonnieres and centerpieces.

But what the hell? A foot in the door is better than one in the ass.

I fill out the application on my phone, hit submit, and slug back the last of my Summit.

A couple days later I'm at AWP, the annual Associated Writing Programs conference, which happens to be in Minneapolis this year. Anywhere between ten and fifteen thousand writers, teachers, and academic types attend, most of whom are also vying to do readings or panel presentations. This year I somehow managed to snag one of those lucky slots, and after presenting in front of a packed room of hundreds of my peers, I drive past a restaurant in my neighborhood with a HELP WANTED sign in the window, and since I'm dressed nicer than usual because of the conference, I pop in and apply for a job.

And, just like that, I am hired on the spot to be a part-time bartender.

I know it's not going to be anything compared to what's going on at Surly, but I figure it's better than nothing, and I hadn't heard anything from Surly.

Later the same day, I get an email from Autumn, the HR person at Surly.

It's full of exclamation points and casual *Heys*, and she apologizes for the late notice but wants to know am I available for an interview tomorrow.

I mix myself a Tom Collins ("practice" for my bartending gig) and fret over my reply. On the one hand, I'd love a job at Surly. On the other hand, the position is for "Event Center Staff—Seasonal." And I just got a bartending job, which I am entirely certain is not what I'd get to do at Surly.

I decide to take the high road and write back to Autumn:

Ack! I'm afraid I've just taken a bartending position at this new place on a lake in St. Paul. I'm super torn—it hasn't started yet and I really have consistently been struck by the tremendous level of service at Surly every time I've been there and would truly love to join your team. If you feel like the position at Surly wouldn't eclipse the bartending gig, I would abso-

lutely love to meet with you and I am available tomorrow, but your time is valuable and I would hate to waste a moment of it. Thank you for your interest and your reply. I look forward to hearing what you think.

All the best,
Matt

I figure that's that. She will probably just delete my email and that'll be the end. I imagine Surly to be a big enough company by now that her job must be kind of like Laverne or Shirley's, where she has hundreds of thousands of applications bursting into her email like they have bottles of beer flying by all day, every day. But maybe not.

She replies within moments:

Hey! I'm a St. Paul person and our whole family is thrilled about the new place—if we're thinking of the same place LOL! Wouldn't want to steal you from them, but it doesn't hurt to talk. No doesn't mean never in either direction. Does any time tomorrow work?

Damn. She called my bluff. I don't know. It wasn't really a bluff. But I just kind of wanted to be let off the hook. But what the hell? I frankly love her tone. "It doesn't hurt to talk"! "No doesn't mean never"! She makes me feel like I'm being scooped by a headhunter, but in a good way.

We set up an appointment for the next afternoon at one, and she says that Dan, the manager of the restaurant, might sit in.

Both of those facts surprise me. The job posting I saw didn't say anything about a restaurant job. I check the website again. Even the banquet staff position isn't listed anymore. What the hell do I know? I figure I'll just go with an open mind.

The other thing that surprises me is that they're doing the interview right smack in the middle of their lunch rush. That seems insane to me for a restaurant, but what do I know?

The next day when I get to the brewery the parking lot is so packed I have to park on the street a few blocks away. I'm still early enough that it doesn't matter much, and, once inside, the frenetically busy host directs me upstairs. "Autumn's already gone," she says, collecting a stack of menus for the group waiting behind me, "but Andrea ought to be able to talk to you. Anyway, just go upstairs."

As I head up, confused, I check my phone to see if I screwed up the time, but I haven't. I wonder what the hell as I scan the upstairs balcony where there's an empty, glass-walled room; a couple of construction guys on a scissor lift are installing banners on the ceiling, and past them, a bank of cooks are all working behind a counter with their heads down. Opposite them at the bar is a woman working on a laptop, and I make my way around a temporary stanchion and say hi and ask if she's Andrea.

"Matt!" she says. "No, I'm Autumn," and I'm so confused, but she springs up like she's going to give me a bear hug but visibly restrains herself and gives me the warmest, two-handed handshake I've ever received.

But then it's all business as she directs me to an empty table in an empty banquet hall and invites me to sit with my back to the rest of the room as she goes to track down Andrea and Dan.

Too jittery today and too nervous in general to sit with my back to such a vast, open space, I decide to stand and look through the floor-to-ceiling glass windows at the brewery below, wishing I had a sample of their finest to calm me down. Within a minute, in walks a guy probably five or ten years younger than I am wearing jeans and one of those ubiquitous, short sleeved, snap button brewer's shirts. He's got an easy, loose gait for a restaurant person (they usually all haul ass everywhere they go, as though their section has just been double-sat and is also on fire), and despite his Pat Riley slicked-back hair and stylish stubble beard, he's got a relaxed, calm demeanor as though he's not really looking for anything in particular in me or this room even, but instead he's just out on his daily constitutional in his Chuck Taylors.

"Hey," he says, much cooler and restrained than Autumn. "I'm Dan."

Andrea, apparently, is still being tracked down, so I nervously chit-chat with Dan for some reason about liquor laws in Utah, and I prattle on about how many glasses you'd need to set down in front of somebody there who ordered a Long Island Iced Tea (five, because Utah laws at the time insisted that no drink could consist of more than one ounce of alcohol and a proper LIIT has a whole bunch of booze in it. And Coke).

After a few awkward minutes, Autumn reappears and introduces Andrea, who's got the flounciest, bounciest curly blond hair of anybody short of Shirley Temple.

"Thanks for letting me sit in on your interview," she says and then tells Dan thanks as well.

He nods noncommittally but with enough of a smile that we both know he's saying you're welcome but not in a gregarious way.

I'm confused. I don't really understand who Dan is. I thought he was the one who was sitting in on the interview. Something appears to have changed overnight, and I'm not sure what. I figure I had just better stay on my feet, but suddenly my lack of preparation and plan to just listen isn't feeling like it's going to be enough to land a job here.

"So," Dan starts, folding his hands on the table between us. "Why do you want to work in a restaurant after so many years away?"

I worry that this is the only question I'll get—that he's seen the nearly decade-long gap between now and my last restaurant job—but I tell him the first thing that pops into my head. "About a month ago, I was at my own birthday party," I say, "and at one point I noticed I had a couple of glasses and a handful of plates in my hands and none of them were mine and somebody hollered at me to stop taking care of everybody and just sit the hell down and have a good time."

Dan and Andrea exchange knowing looks and smile.

"It's just kind of never left me," I say. "Taking care of people. Guests in my home, guests in a restaurant. It's all kind of the same thing to me. I can't help myself. I love it."

Dan nods and kind of looks to Andrea to see if she has any questions, but she demurs again. He unfolds his hands and leans in across the table like he's going to let me in on a secret. "So this restaurant up here—maybe you've heard—it's going to be a pretty adventurous, fairly high-end kind of concept. You know," he says, "like duck tongues and shit."

"Mm," I say, laughing noncommittally, "duck tongues."

He does this smirk-laugh thing that's fairly inscrutable, and I have no idea if he just used duck tongues as the most preposterous example of what they might do or if it is an actual sample of what they're planning. He then asks another question about pairing beer with fine food, and I tell him how Wendy, the last sommelier I worked with in Utah, was really progressive and worked hard to pair not just wine but also beer with food.

He seems a little dissatisfied with this, picking some lint from the knee of his jeans, but then follows up. "Do you drink?" he asks. "What's your favorite Surly beer?"

I'm a bit thrown by the first question and hadn't even thought to have something sophisticated to say to the second, but I wager on just trying to be honest. "I hope this doesn't sound like brownnosing or whatever," I say, "but I think Surly Furious is simply a perfect, unimprovable beer. It's like a tuxedo—a tuxedo of beers."

I go on a bit about how and why, but Dan stops me. "So what food would you pair with it?"

This, I feel, is the real interview right here. This one question. I panic. I have no idea what to say. I like Furious all the time, with everything. It's fairly expensive at the liquor store at ten bucks for a four-pack, but I still usually get it at least once a week. I try to think of the most sophisticated food I've ever served, but all I can think of are expensive filets mignon or cedar-planked salmon, neither of which feels like it compares at all with duck tongues and the like.

But I've got to plant my flag on whatever ground I have. "Well," I start, "it's a pretty big beer, hop-forward, lots of citrus. Really floral. You could go one of two ways. You could try to match its flavors with something big and bold—something Cajun inspired, say, like red beans and rice or hoppin' John or étouffée." I'm only able to think of the foods they're going to be serving at the lakeside place, and I'm rushed with a feeling of betrayal, but then, as sweat starts to trickle down the inside of my shirt, I remember I've got to finish my answer. "Or you could go the other way," I say, "and pair it with something with more delicate flavors that the flowery, grapefruity flavors would pull out. Something like a blackened mahi-mahi."

Despite the fact that there's nothing "delicate" about blackened mahi-mahi, Dan gives me a tight-lipped smile and says that's pretty much all he has for me.

In a kind of pro forma way, Andrea asks me a few questions about banquet staff, but I'm so frazzled from the original switch-up in the interview that I can barely stammer out any kind of coherent reply. They both seem to be eager to get back to whatever it was they were doing and disappear behind banquet hall doors, leaving me alone again in this empty room.

Thank God I've already got the lakeside gig, I think as I walk to my car, because Lordy did I muck this one up.

The next Monday I'm at the climbing gym in Minneapolis, and I take a break between routes to check my phone, eager to hear back from an editor at *Outside* magazine who's got a piece I wrote about ultra-running. I've been trying for years to get *Outside* to take something of mine to precisely zero avail, but I've heard they pay like two dollars per word, so if they took my 10K-word essay I could make enough to get through the rest of the summer and then some if these restaurant jobs don't pan out. When I get to my locker and see a missed call from a number I don't recognize, my gut does a couple of forward handsprings as I begin to play the message.

It's not Alex at *Outside*. It's Dan at Surly.

I halfway expect him to tell me how disappointed he was in my interview and that it was frankly offensive and stuntish of me even to have presumed to be qualified to apply, never mind actually interview, for the job (dismissing the fact that I didn't actually apply for the job in question), but instead the message says he wants to discuss my coming on board.

I'm still wearing my harness and climbing shoes, and I endeavor to not be the kind of guy who makes phone calls from the locker room. I decide to play it cool and call him back after I'm finished at the gym, but when I'm back in my car, my stupid hands-free car stereo won't link up with my phone, so I quickly pull over, turn the car off, and call him back, but it goes right to voice mail.

Now it's getting on toward half past four, and I still need to get to the grocery store, pick my kid up from aftercare at his school, walk the dog, make dinner, and then get to the four-hour scheduled orientation for the lakeside job.

The traffic is heavy and capricious on Franklin Avenue in Minneapolis, and I'm afraid there's slim to no chance that I can easily pull over if need be, but Dan doesn't call back by the time I reach the Seward Co-op. I make sure I have my phone and run in, dashing to get the few ingredients for Marcella Hazan's simple but sublime pasta and red sauce recipe, but something tells me to check my phone and it's ringing and it's Dan.

I panic for a second, wondering if I should drop my basket and run for the door, but there's just no way, and I don't feel like I can let the call go to voice mail again, so I answer as calmly and jovially as I can and tell him where I am and ask if he can hear me.

"Not really," he says. "Just give me a call back when you've got a better signal."

Shit, I think. I sound like I'm too busy to take his call or be available when he calls. I know it probably sounds stupid to people with important jobs, but in the sunny prospect of a new job at a restaurant, everything counts, foremost punctuality, enthusiasm, and simple availability. Nobody wants to hire a waiter who's going to be late, difficult to get ahold of, or tedious to work with. All other skills in this industry depend on those three, of that much I am certain.

When I get back to my car with my groceries, I leave the engine off and the windows up, despite the fact that it's a fairly hot, sunny May day, and I return Dan's call, but the stupid thing goes straight to voice mail again.

I nervously check my phone about once a minute for the next couple of hours, but no luck.

The next day I'm getting ready to go back to the lake place for more training when Dan calls back. He sounds unfazed by the frivolity of what it took just to make contact.

"So," he says, "I've pretty well got my team set, but what I don't have is anybody to pick up the slack if somebody needs a shift off or gets sick or whatever, so what would you think about coming to work at Surly on Wednesdays and Saturdays?"

I like the guys I've met at the lake place so far, but things feel ramshackle already, and though we're supposed to open soon I still don't have any idea when or how often I'll work there. I think about how damned busy Surly always is and how much I like everything they do and how impressed I have been with the service every time I've been there. I also think about how I don't really want to work full-time at the lake place, waiting on neighbor after neighbor, night after night. I don't know. I might love it. I might hate it. I also might suck at it.

But then I also might not ever really be able to afford to go to this restaurant at Surly. I'm afraid that working at either Surly or the lakeside place will kind of wreck dining at either of them, but working just part-time at both might leave enough of the novelty of the experience, I hope.

What puts me over the edge is the thought that I bet they are absolutely killing it tip-wise at Surly. I'm not positive, of course, but I think I stand to make more than two hundred dollars per night. A simple reflection on the fact that two hundred dollars a day is more than what I currently make in academia—more than twice what I'm earning on sabbatical—seals the deal.

"Hell yes!" I tell Dan, not sure if my answer is quite parallel to his question.

"Sweet," he says. "I'll shoot you an email to get your availability for training."

And with that, I head out the door for my second day of training at the lake place, feeling thrilled with both of these great opportunities—either of them probably better than any restaurant job I've ever had—but also worried as a rookie polygamist, fearful that I'm kind of forsaking both wives instead of choosing just one.

Chapter Two

CURING THE HEART

ARRIVING AT QUARTER TO NINE AT A BREWERY ON A MON-
day morning feels weird. I park on the street and walk to the front door,
but they're locked tight and there's no sign of life inside, and I panic,
thinking that my orientation is supposed to be elsewhere or at another
time, and I pull out my phone and begin to look for the email from
the HR person when a young woman with a closely shorn head comes
around the corner.

"You here for the orientation?" I ask.

She can't be much more than twenty or so, and she's so small and
skinny I bet she barely outweighs an empty dry cleaning bag. She has a
cagey but also sheepish look about her, as though she's scared but trying
to come off as tough instead. She's either fresh from a stint in a military
school or a punk rock band or both.

"Yeah," she says. "I'm Tina. And I'm kind of freaked out that this
is the wrong place, like we're supposed to be at the other brewery—
Brooklyn Park or wherever."

Just then a woman comes around the corner. She's probably in her
early thirties and radiates toughness and attitude, but in a good way.
Like she knows how to plumb a sink or win in a knife fight.

"What's up? I'm Remi," she says, firmly as a pipefitter. "How the
fuck are we gonna get into this place?"

But she doesn't so much ask it like a question but rather like a riddle
she's going to solve, and with that she dives through some shrubbery
and into the beer garden on the other side. I don't believe there is much
that is going to stop this woman from getting in wherever she wants.

Tina and I look at each other like we're going to let Remi do what Remi is going to do. If this is a first test of some kind—like we've got to make it to the orientation on time no matter if the building is locked or not—at least Tina and I have each other to vouch for the fact that we were early and didn't do any damage to the building or the landscaping in our attempt to get inside.

While we wait, I take a moment to consider my surroundings. The Surly destination brewery is only months old and sits, literally, on the industrial wasteland between Saint Paul and Minneapolis near Prospect Park. Surrounding the huge two- or three-acre plot are towering disused grain bins from when this part of town was ground zero for processing and shipping all manner of cereal grains by train to the rest of the country. Pillsbury and General Mills, both legacy Minnesota companies, began, after all, as purveyors of grain rather than how we think of them now—more as makers of bags of flour and colon-clearing cereals. The ten-story cement grain elevators stand like oversized monuments to themselves, hollow and abandoned but impossible to get rid of. One long row of these elevators stands several stories above the rest, the words UNITED CRUSHERS graffitied along the top. It's baffling to think how anybody got up there, never mind how the artist then dangled over the edge to paint each story-high letter.

The Surly building is a work of shiny aluminum, dark grays and greens with pops of red wherever its name is, and its situation in this part of town feels like an update of the place but also considerate of its surroundings. Stainless steel grain silos shaped like a three-story six-pack stand sentry right at the front of the facility. Rather than cover up or gloss over what the place is, Surly has put it front and center as though to remind everyone from the very first that this is what it's all about: beer.

As I ponder the style of its hard-edged, sharp-angled architecture (I heard someone say it's called "brutalist" design), a shortish man with a smart goatee and hard-soled cycling cleats opens the door from the inside. He wears glasses and looks somehow down and through them as though to exaggerate his scrutiny like a small and more exacting Dumbledore.

"Gary," he says as he struggles to hold the door open for me and Tina. "Nichole is waiting for you upstairs, or at least she should be."

"Hey, Gary!" I say exuberantly—too much so, in fact—and Gary kind of fake smirks at me as I slip through the door and wait for him and Tina and then Remi, who emerges from the shrubbery like a feral cat.

"I'll figure out your names later," he says. "Let's get you going."

I can tell that Gary's the kind of guy who likes to lead, or at the very least to be in firm control, and that our being locked outside on our first day, waiting like sheep, has got him frazzled and irritated—all the more so because he clearly has also just arrived and therefore hasn't had a chance to change out of his cycling shoes, which makes him look unprepared, the last thing in the world I gather he wants anybody to think of him. He clip-clops his way ahead, and first we head straight toward the gleaming stainless steel brewing fermenter, and it's framed perfectly by an entire wall of glass. The floor shines impeccably, and the steel of the tank glitters as though it's a symbol of what it does rather than the thing itself. It's inspiring and moving in an oddly industrial and beery way.

Gary leads us clickily up a stairway by the gift shop, and I make the jokey, nervous comment that this is the first time I've been here when the entire beer hall wasn't packed. Gary doesn't turn or reply but seems to like being on the receiving end of comments he has no intention of returning. It feels like a conspicuous power move, but somehow I can't hold it against Gary. Restaurants can be rife with frivolous people like me who are, trying to make up for it, overly eager to please.

As downstairs, upstairs is quiet. The dining room of the restaurant now has tables and chairs—lovely gray cloth chairs and dark, solid-wood tables—and museum-quality red and clear glass pendant lights hanging from the now cloth-baffled ceiling. The first wall to the right, as below, is still all glass and definitely lends the impression that this place too is all about beer, but here a kitchen completely unobscured from view—at least from the waist up—guarantees that food will be at least an equal priority. I imagine chefs can go their entire lives without a chance to work in such a vaunted space as this.

The open kitchen is quiet now, but its stainless steel hoods and walls sparkle beneath hidden, harsh lights. Everything looks and, likely, is new and spotless. The only signs of work having already been done are those clear plastic Cambro containers on shelves with dry ingredients in them, blue painter's tape and Sharpies labeling everything. Gleaming stainless steel pots and pans are nested neatly in stacks, as are black and gray plates. Empty squeeze bottles are lined up, patient and ready as novitiates. And then, as much of a surprise as it would be to find Waldo standing back there, on the top shelf in full view for anyone to see is a large plastic tub of nuclear orange Utz Cheese Balls.

On the near side of the room is a wood-planked bar with a huge twenty-tap system they're apparently still installing—plastic hoses sprout everywhere—but on the far outside wall is nothing but floor-to-ceiling glass, and even though the view is of the grain silos and, beyond it, the Minneapolis skyline, it instantly becomes one of my favorites. Though I've never been on one, it feels like the high, broad bridge of a mile-long supertanker, and I can't wait to see where this all will lead.

Having spent the better part of the past twenty years trapped in largely windowless, square Eisenhower-era classrooms with that smell of old, damp carpet and chalk, I'm freaking stoked to be here. Compared to any given college classroom, this is a museum—an art gallery—an example of modern, urban art in and of itself. Even, maybe especially, the cheese balls.

Gary leads us past the bar and around a corner and into a private dining room that has a long, lozenge-shaped hardwood table apparently made for this space and then ducks out, presumably to put on more sensible shoes. At each place setting a red folder awaits, and at the end of the room, a happy young woman with great kinky hair is setting up a laptop.

"Hey! You guys made it! I'm Nichole! What's up!"

She gets up and dances over to shake each of our hands, happy as a woman at her own birthday party rather than an HR person preparing for a PowerPoint presentation first thing on a Monday morning. It is not possible to dislike this person.

There are probably twenty seats at this long, officious table, but by the time the clock hits nine, it's only the three of us and Nichole, so we wait, nervously not talking for a few more minutes, pretending there's urgent matter on our phones.

"I can't believe there's only three of y'all," Nichole says, satisfied enough with her computer's readiness and the hour to begin.

But just then another three people come in, all but one of them annoyed and/or hungover enough to be irritated by a 9:00 a.m. orientation and a building they couldn't get into and therefore seeming to be late on their first day.

"Hey, cool," Nichole says, "you guys made it—you passed the first test! Getting into the building!"

Another manager could make that kind of comment seem unacceptable or passive-aggressive, but Nichole comes across as just trying to downplay our first-day anxieties, and it feels palpably clear that this will not be held against any of us.

"So let's start by—I don't know," Nichole says, "getting to know each other by saying not the typical stuff maybe but more like—I don't know—how about something surprising about each of you."

I suddenly am so nervous I don't even hear what Nichole says about herself other than she's Autumn's assistant and Autumn's got a sick kid this morning or she'd for sure be here. I'm sitting closest to her so I go next.

"Hi," I say, my voice almost audibly breaking. "I'm Matt and, other than this, I teach—I'm a teacher—I teach . . . I teach creative writing," I say as though asking a question rather than stating my so-called career. "And for something surprising—that was the second question, right? I'm actually opening two restaurants in the same month."

I meant what I said to be impressive in a funny, wacky, who-would-do-that kind of way, but nobody seems to take it in any way other than I'm a prick. I wish I could take it back and say something—anything—else. Tell them that I've never tasted Red Bull, that I've run fifty miles in one day, that I have a seven-year-old whose favorite foods are sushi and ribs, that I've been happily married for fifteen years. But there will be no do-overs this morning.

Tina goes next and says something about one of the Dakotas and how she works at Pizza Lucé and then says something self-depreciatory about how there might have been too much "Jaymo" last night. Nobody seems to get her lingo, and she blushes roughly the shade of an heirloom tomato. "Jameson," she says. "Irish whisky? Hello?" She seems very young, aggressive, and stressed out, like an up-and-coming Sinead O'Connor before she understands what it means that she's Sinead O'Connor.

Remi goes next and tells us she's from La Crosse, Wisconsin, has just moved here, knows her way around beer from a decade of bartending and running a liquor store. She projects the best kind of confidence—like she full well expects you will like her, and she knows she's right. And she is, and we all do.

"I'm Sofia," the next woman says. She's got curly, uncontrollable hair like me, and even though she's only said two words I can tell she knows her shit. She rattles off three or four restaurants where she's either served or managed. At least two of them are the kinds of restaurants I've recently waited for over an hour to eat at. She has a breathless quality to the way she talks, as though if she pauses she might not be able to start again, but she also seems kind and full of good humor and self-aware enough to be striving to contain herself. I am willing to bet she would be the absolute life of the party with just one or two drinks in her. Big trouble perhaps after three or four, but in a good way.

The next woman across the table pauses, considers her words and her demeanor. She's wearing a black sweatshirt with the hood up and a black skirt with black leggings, and I get the immediate impression that this is more than just what she happened to grab this morning. It's who she is. The Dark.

"I'm Rae," she says, pausing dramatically to lower her cowl. She sighs as though it takes great effort to deign us with her work history. She too names a couple of places I like to frequent, including a super popular spot my friends and colleagues tend to frequent after literary events in Minneapolis. "Where do you teach?" she says, fixing her gaze on me.

Her question catches me visibly off guard. I had decidedly not wanted to bring up the name of my school. It undeniably has something of a snobby, rich-kid reputation. Almost all the buildings on campus have facades cut from the same limestone, everybody walks only on the sidewalks (never on the meticulously kept lawns), and between the buildings and the landscaping and the students who tend to groom themselves in strict accordance with whatever's trending (Ugg boots and North Face puffy coats at the moment for the women; unfortunately, sweat pants and Hollister hoodies for the men) it can feel pretty Stepfordy. I am grateful for my position in higher education but not all that eager to put my school's image in line with my own.

But I come clean and say the name.

"Oh," Rae says, unimpressed. "I'm in your grad program. I'm probably dropping out though."

"What?!" I say, but there's no time, and everybody laughs nervously not knowing what to say about grad school or the two of us, and we're far enough apart that it's clearly a conversation that's going to have to wait for later.

The last guy announces himself as Hans. He's probably thirty years old, but he's got a blond, Teutonic beard that makes him seem both younger and older, and his arm-slung-over-the-back-of-his-chair attitude and his backward ball cap and decisive lack of a smile announce that he's only half in—whatever the question or commitment is—and he says he's also a server at some restaurant we've never heard of but he clearly thinks we all should have.

I ask where again.

"Brasserie Zentral!" he says, as frustrated as Putin being asked what was the name of his country again.

Even though we've only just met, I get the clear impression that I've got some work to do if I want Hans, Rae, and Tina to like me, and I feel foolish for the thought, but it's real. I really want these guys to like me. Remi and Sofia are both already on everybody's side, and I'm so grateful for them.

My enthusiasm for the place is still fresh, but I am riddled with doubt as to whether I have the skills or the personality to match up with these serious restaurant vets. None of them appear to know each other, but it feels like they're only one person—two at most—removed from one another. They are their own people and they all know it. Even little Tina Jaymo is in. Me, I'm at best an older version of them, but one who has been conspicuously out of the game for quite a while, and given that I've never played it in this state seems to make me suspicious—an interloper at best, a poser freak at worst. I didn't expect to feel so old or so out of place, but I do, and I don't like it and am not sure there's much of anything I can do about it.

But then Nichole takes the reins and begins orientation, and given that she's at the short end of the room, opposite the rest of us new folks, I can concentrate entirely on her radiant, positive energy as she begins to tell us the story of the founder of Surly, Omar Ansari.

* * *

If ever there was a self-made man, Omar is it. Or at least his parents were. His mother was born in Germany and his father hails from Pakistan, but somehow they met and married and moved to America and had a son, Omar. They came to this country with almost literally nothing, but his father worked his ass off in industrial jobs and eventually managed to found his own abrasives company, which he hoped to pass on to his son. Things veered toward a different course when, in 2004, in their family driveway in Brooklyn Center, Minnesota, Omar and his brewer buddy Todd stumbled on a batch of what was destined to become the first of Surly beers.

Demand for Omar's beer quickly outstripped the brewing capacity of the family driveway and garage, and his father eventually gave his son a space in the corner of his abrasives facility to brew more and more beer, and before long it became clear that they had something with real potential on their hands.

It all began in an effort to simply make beer that didn't suck. Throughout the eighties and nineties, the beer scene in the Midwest was grim: pretty much nothing but skunky, "lite" lagers from industry stalwarts that have been making the same weak beer for generations. Schell's, Hamm's, Grain Belt, and, across the border in Wisconsin, Leinenkugels, Pabst, Miller—no great shakes. There were choices, sure, but choices without distinction. Compared to the beer they brewed in Belgium, Germany, Ireland, even England, it made Omar angry how most people in the Midwest settled for shitty beer. Omar's attitude about beer before he started to brew his own was downright surly, and hey, he thought, that's what I'll call it. Surly.

Originally available only in kegs that Omar would sell one at a time out of the back of his truck to bars around the Twin Cities, Surly beers before long were available in cans at liquor stores throughout the state. And when at last it became clear that demand was only going up and they had reached a physical, literal wall, they began to seek out places not just to make another canning facility but rather to create what came to be called this destination brewery and beer hall.

* * *

I first encountered the beer hall at Surly a few months earlier with my wife, Jenae. A host showed us to a seat at one of the long, family-style tables, and it was just so busy and solid with diners and drinkers that I expected it would be hours before we'd see our server, but before we even got our coats all the way off a beaming young woman warmly greeted us and introduced herself as Cate.

As both Jenae and I have done plenty of time in the service industry, we are always grateful to servers who do their jobs well, and Cate shot straight up the ranks as she introduced us to the place and the menu and continued to do so in every regard throughout the meal. But so did all of the support staff. Everybody was competitively kind and knowledgeable and attentive to every detail, making sure our water glasses were always full, getting us samples of any beer we wanted to try, telling us all about what was on our extravagant charcuterie board, and giving us the details about our miraculous "hog frites," which was something like an American poutine of smoked pulled pork, French fries, spicy giardiniera, and melted cheese.

I couldn't believe how good it all was. Nor could I believe how straightforwardly they put it, actually in print, right there at the top of the menu:

Make great beer.
Have a great time.
Give a damn about your community.
Be independent.
Don't be a dick.

I was so smitten, I came back the very next day. I was intrigued to see if it was a rogue experience or if it was actually systemic. I had to know where all this was coming from, and when the friend I was with headed for the bathroom, I hailed the bartender and asked him what the deal was.

He gave me one word in reply: Linda.

Until recently, Linda was operating a much-loved restaurant in Linden Hills (pretty much the highest-rent Minneapolis neighborhood there is): Café 28, a cool, quirky, cozy bistro in an old firehouse next to a kooky kids' bookstore with free-roaming cats and chickens and even a mini-door for kids only. Todd, Omar's old friend and now Surly's head brewer, is her husband. Chris, the bartender I was talking to, said he also worked for her somewhere I didn't recognize. Pretty much everything great about Surly, Chris argued, came not just from Omar but also from Linda and Todd. "Especially Linda," Chris said, reminding me of that cringey way Hamlet was a little too jealous of his uncle (his father's murder notwithstanding, of course).

In my orientation, I continue to see evidence of Linda's style throughout the day in the steps of service and the company ethos and even the mini-manifesto on the menu: they're all no bullshit and jargon free. Nothing feels like it's been vetted by a lawyer or a seminarian. There are far too many examples to keep track of during the training, but phrases like "being sick sucks but it happens—don't come to work contagious" and "work hard and you will be recognized for your asskickery" are going to stay in my head for a long time.

I love that her model is full of trust, enthusiasm, empowerment (which is already too jargony a word, I imagine, for her, but I can't help myself), and, above all, a sense of humility and humor. It's clear that the top three priorities at Surly via Linda are (1) people; (2) beer; (3) food.

In the few months since Surly opened their doors, they've been slammed almost all the time and have rocketed to the front of both the culinary and the beer scenes in Minnesota, and with that comes both more celebrated notoriety and the implicit risk of losing some street cred. Whenever somebody or some enterprise goes beyond its roots to the next level, there's always going to be some cabal of schmucks who say they're not "keeping it real" anymore. As though keeping it real implicitly means selling your plasma, scrounging change for bus fare, and splitting packages of ramen with your bandmates instead of getting paid what you deserve.

Yes, Surly is successful—there's no denying that—but they're doing it their own way, I thought. Sure, this was orientation and they were serving their own Kool-Aid by the gallon, but I was thirsty and just drank it up.

By this point in the presentation, it's almost noon, and it's clear that things are about to wrap up. "We've got a pretty good thing going," Nichole says, pausing before she advances to what will be the last slide of the morning. She clicks ahead without looking and reads the red words as they shimmer onto the black screen:

Don't fuck it up.

She asks if there are any questions, and though I can tell that most of the room is over the spiel, I can't help myself and raise my hand.

"Yeah, Matt?"

"How is it even possible that anyone can ever use PowerPoint again after that bomb of a slide?" I say. "I mean, you may have just broken PowerPoint for good."

Just thinking of using anything approaching that slide within the austere halls of academia is enough, I imagine, to bring on a murder of university lawyers.

"I don't know," she says, "but we're just going to have to deal."

Though lunch is a most welcome break, I fear that I've reached critical mass in terms of how much data I can process and I'm internally super conflicted. Frankie, my Australian shepherd, has been alone in her crate since 8:30, and if this training goes until 5:00 it will have been the longest day she's ever spent there, and that's simply not okay with me but it's my first day here and I know I can't beg off early.

I sit next to Hans, who orders the charcuterie board, and Rae. Both she and I order the "cold fried chicken," and while we wait I try to get from her why she's dropping out of our program, but she ignores my question and I get the feeling that conversation will have to wait until she decides who I really am. A professor/dick or a waiter/dick or something else/dick.

Somebody at our table ordered the brisket, and they pass it around to share and I take a forkful as it passes by. "My God," I say, "I think I like that more than my wife."

"Really?" Rae says archly. "More than the brisket she makes or more than your wife herself?"

I was just trying to be funny, but I recognize I have just said something stupid and vaguely sexist and I've been called out. I want to elaborate that I do almost all of the cooking at home—especially the brisket—and lots of the cleaning and my fair share of the dog walking and diaper changing, but I know anything I say will continue to be held against me.

Jenae would understand, I believe. We've been together for more than twenty years, married for more than fifteen of them. She knows what a dumbass I can be and wouldn't be surprised at all.

"I love you," I'd say on the heels of such a stupid comment, all apologies.

"Yeah, yeah, yeah," she'd say right back.

Regardless, Rae won't be my pal for quite a while, I'm afraid.

Fortunately, lunch comes and goes quickly, and after a solid meal I think I have enough in my system to make it for another couple of hours, but when we get back in the room Gary is there, and though Nichole starts, once the subject tilts toward beer it's clear that Gary will allow no other to intervene. He is not a beer drinker, or a beer enthusiast, or a beer maker. Gary is a beer scientist, and this, other than the lab, is his element. Gone is the peevishness from this morning. Now he stands at the head of the table, bogarting without apology Nichole's training dais.

Gary's demeanor is that of a particle physicist who has chosen to apply his doctoral degrees toward the production of beer—which he considers not at all beneath himself or science—but with that comes the necessity of dumbing down string theory and the science of beer to a bunch of waiters who are quite possibly still drunk/hungover from last night, hence, their inability to court the fine nuances of fermions and quarks and their throbbing need to just get the basics. I imagine Gary, five hundred years ago, dressed in the garbs of a cleric or a priest, trying to describe the methods and the philosophy behind alchemy, when all anybody he talks to cares about is how shiny the gold is.

I like Gary, but Gary, I'm pretty sure, doesn't like us. But then after the official training of the day is over (mercifully only an hour later), Gary leads us back down the stairs in a kind of reverse of how the day began and articulates how much of my experience of the visual impact of the place was absolutely intentional. It turns out that the architects and developers of the building are in fact the same ones who designed the iconic Walker Art Center in Minneapolis. Every single detail here—

from the view off the balcony to the position over an aquifer that pro-
vides this brewery with water of the same pH as the original brewery
fifteen miles away to the ramped shape of the tops of the second-floor
rails meant to both imitate the profile of the brewery itself and prevent
guests from putting their glasses on it and courting what would be an
ugly spill—was designed to have the impact of experiencing a work of
art. It makes me very happy.

Gary takes us back outside and around the back of the building. Past
the loading docks and a couple of spent grain hoppers is the employee
entrance, but instead of leading us there he takes us to the edge of the
vast parking lot, which is as big as two or three football fields.

"I didn't take you out here to kill you," Gary says, but we don't laugh
as the joke seems too forced. "Actually," he continues, pointing beyond
the stacks of shipping containers and the stands of tall weeds toward
the limit of the property where a couple of boxes stand in the distance,
"I wanted to show you those."

"Beehives!" Rae says, the first glimmer of life in her voice all day.

"Beehives," Gary says, also as gleeful as he gets—which is still not
that gleeful, but the uptick in tone is notable. "And we're planting wild
clover for them to pollinate. We won't be able to use the honey or anything until next year, but cool, right?"

I've been stung too many times to like bees all that much, but the
thought of a place in the middle of the Twin Cities establishing a colony
and planting clover so that they can harvest a little honey in another ten
or twelve months—Gary's right. It is pretty cool.

I spend some time that night reading through the stuff in the packet,
expecting there was a lot more to it, but in fact our first-day orientation
was pretty thorough. The other half of the packet is all descriptions of
the menu—super detailed, single-spaced descriptions for each of the
twenty dishes. I figure tomorrow's training will cover it all, so I don't
waste my time with it and try to turn in early.

But when I arrive at Surly the next day at noon, I'm barely on time
as the place is absolutely packed with lunchtime diners, and when I get
upstairs almost everybody from yesterday is already there, sitting in the
dining room against a long bank of tables, but our numbers have more

than doubled, and I recognize several of the faces of servers and bartenders from the beer hall. Taking in the full staff for the first time, it's clear that this is not a batch of amateurs. These people are pros, I think, and I take one of the only available seats across from Sofia. Cate, the server who waited on Jenae and me, sits at the seat next to me and gives me a warm smile and then says, "Hey, mama!" to one of the other servers down the table.

Everybody is mostly quiet as they either look over their menu descriptions or flip through index cards. One woman a couple of seats away has a stack of what appear to be custom colored and cut cards, covered with writing that is tiny and inscrutable but clearly meticulous, and the floor of my stomach drops a few inches like an elevator whose cables are about to snap. Apparently, not only were we supposed to go over everything last night, but we were supposed to *study* it.

I am fucked.

Dan stands in the middle of the dining room right at noon and welcomes us back. He's wearing a Surly brewer's shirt and Chuck Taylors again, and he looks a little too casual for the occasion, but his demeanor is all business except for his inside joke of a grin.

"So," he begins, "don't forget that you guys are getting paid for your training, so take it seriously—it is, in fact, work. I clocked you all in, but when you leave today, make sure to clock yourselves out. Also, Gary dropped these by." He hands out a dense packet, each page with an elaborate description of a single kind of Surly beer, replete with the hops, malts, yeasts, colors, and ABVs. "Don't fucking lose these. This stuff is literally trade secret information, available nowhere, to no one, except here and now to you."

Once again, he has us introduce ourselves, and this time he wants us to talk about why we're here, and I have no idea what he means. Therefore, of course, he decides to start at my end of the table.

Only two people go before me: Hans, who is still aloof and disinterested but also clearly as professional as it gets, and Cate. None of what they say sticks because I'm whirling through what I'm going to say before I say it, but then there I am standing in front of all of them and it might be noon but it feels like most of these guys have just gotten up and they look, suddenly, neither particularly interested in what I have to say nor intrigued to learn why I'm here, and even though the last thing I wanted to do is again announce myself as anything other than somebody who is excited to be here, I find myself rambling. "I've waited tables for a bunch of years, but I'm a teacher," I say, "and I have student loans that—to paraphrase Bruce Springsteen—no honest man can repay."

No one laughs, and I suddenly feel like I may as well have just quoted Don Ho or Lawrence Welk it makes me feel so geriatric—they're not too young to know who Bruce freaking Springsteen is, are they?—but instead of letting it go I ramble on, saying something stupid and immediately regrettable about how I wasn't necessarily even looking to get back in the service industry but I've had such a kick-ass experience every time I've been to the beer hall that I just couldn't wait to be a part of what was going on and—finally—I sit down and shut up.

My ears are burning for the next few minutes so that I only zone back in and hear what the last few of them are saying to understand that what Dan had apparently really wanted to know was what qualifications we bring to the table that will make us an effective team. All of them, of course, have worked at great and busy restaurants in Minneapolis, and several are advancing toward becoming sommeliers and, the beer equivalent, cicerones, and more than a few of them brew their own beer as well.

Even though I've technically been hired already, my job here feels as far from a foregone conclusion as possible.

At the end of the table, a woman with long, curly blond hair and iridescent blue metal glasses introduces herself as Linda, and I recognize with a shock that it's The Linda—Linda Haug, the boss of this whole operation. She doesn't smile at all, and she seems as nervous as somebody awaiting oncology results, not at all like the cool mastermind to this multimillion-dollar operation I'd imagined, but there's something nonetheless magnetic about her. She doesn't seem Midwestern. She's emotional but not simpering like most of us, certainly like me. She's funny and awkward but also shrewd and to the point.

As another server with a moustache as big as a bird of prey introduces himself, a guy in a brewer's shirt sits down behind Dan. He's got cheap plastic sunglasses on his head, and I try to read the printing on his shirt to figure out if he's a maintenance guy here to take a look at the plumbing or a brewer on break or what, and I barely resist the impulse to tell him this is a staff training rather than some kind of open meeting.

Another pair of service guys, completely indistinguishable from him, are going around the dining room from table to table with a level, adjusting the feet as needed. In contrast to every other restaurant I've ever worked at or been in where tables are leveled (if they're leveled at all) with matchbooks or soggy bar coasters, this place is having it done by experts. As with almost everything else, I'm impressed, but I wonder why this third dude isn't helping them out. They certainly don't need a supervisor.

"Omar," Dan says, turning to the random maintenance guy, "do you want to say a few words?"

The guy—Omar, apparently—takes the sunglasses from his head and twirls them in a gesture of embarrassment and a little frustration.

I am feeling like such a complete schmuck for not doing my homework now in at least two ways. First, I didn't even read through the menu, never mind study it. Second, I didn't even have the sense to Google what the founder of the company looks like so that when I see him for the first time I don't mistake him for an HVAC repair guy. He's not some random dude. It's the man who pretty much singlehandedly is the reason we're all here at ground zero for the craft beer-and-food movement in Minnesota—arguably the entire Midwest.

"Thanks a lot, Dan," Omar says, but he clearly means to say that Dan is a little fucker for putting him on the spot.

Once again, I'm too rattled by the stupid things I've recently said and by the fact that Omar and Linda and all of the rest of these badass servers are here right in front of me to be able to remember anything particular Omar says. What's most memorable is just the way he's kind of sheepish in front of his own employees and the fact that he doesn't have much at all to say. I probably rambled on about my stupid student loans and Springsteen longer than Omar talked about his own thirty-five-million-dollar facility.

After Omar leaves, a cook on the line says that the first dish will be out in just a minute, and our attention turns to the kitchen. They're all heads down, busily working at dishes we can't see; the sound of something sizzling in a sauté pan and a couple of plates clicking against each other is all the noise coming from these five cooks in white short-sleeved shirts and black aprons. As a group, they move like extensions of one body, without words, without gestures, parrying and dodging one another as an element is added or adjusted or readied onto the plate. All of their attention is on the dish at hand, and they focus on one another's work with all the interest of one master painter studying another's work. As one of them uses large tweezers to place a big round leaf on a plate, a couple of the other chefs nod with approval, as though that is exactly where they would have placed that one leaf.

I have never seen a group of chefs with so much focus, so much attentive and collective coordination. I'm expecting profanity or for one of them to yell that he's waiting for the chicken or another to say that it'll be three minutes before the filet is done, but none of them say anything, and in this way they're more like a choir, all of whom have preordained parts, and they each have the score in front of them and are counting out the measures of rest until their part comes in while also listening with interest in the performances of their colleagues.

The servers and Dan and Linda swarm from the banquette to the chef's counter, and as a plate is set before us, someone mutters, "Oh, the duck tongue, right?"

Miko—the mustachioed one—says "fuck yeah" with great passion and anticipation.

Yet again, I feel like an immense and worthless screwup. Apparently what is happening today is not that we will be talking about the menu. Instead, we were clearly supposed to have studied last night so that we could be prepared for the menu to be presented to us, dish by dish by dish. I figure what the hell and put the packet back in the folder, grab a fork, and decide I may as well just dive in with the rest of the folks and try to play along.

I remember back when Dan interviewed me and asked what I thought about serving high-end food like duck tongue. I thought that was a joke, and yet here it is about to be on my fork.

"And that's the tamarind sauce," somebody says.

"Yeah," another answers. "Tamarind bourbon. And pickled cauliflower."

"Fucking good," yet another says around a bite of food.

I manage to horn in and spear a bite.

My mind reels as the flavors are bright and delicate but still bold. The sauce is slightly sweet and a little sour as well, but also nicely sharp and crisp with a touch of the pickled cauliflower following suit, but best of all is the ever-so-slight crunch of the tempura batter, followed by the again sweet-and-sour flavor of the duck tongue itself. I expect a gritty toughness or an awful chewiness like the time I tried whale blubber, but instead the tongue is more delicate than any meat—or organ or whatever part of food tongue is—I've ever had.

It's not every day that a forty-two-year-old man who has spent nearly a third of that time working in restaurants (the rest of it fairly enthusiastically in pursuit of new and adventurous cuisine) can come across both a food product and a flavor that feel at once comfortable and familiar and yet without predecessor or peer. If this is any foretaste of what is to come with the rest of the menu, I am simply not going to be able to keep up.

I think I'm lucky, therefore, when the next dish comes out and it looks simply like a little slider-size version of a Reuben. A pony-tailed cook lurches out of the kitchen with another pair of these little guys, sets them down, and spins back into the kitchen.

A server named Esme—the one with the multicolored flashcards—begins to slice the thing into bite-sized pieces for everybody. "So," she says. "That's the charcoal bun, the caraway kraut, and is that the fancy sauce—or the Mahon fondue?"

"Yeah," the bald cook who appears to be in charge hollers from the kitchen without looking up. He's got tattoos all over his arms, and he's bald but has a robust black and gray beard. "It's fucking straight-up fancy sauce, Esme. Thought it'd be funny."

I don't know what he means or what fancy sauce is or why it's a joke. Neither he nor Esme appears to be thinking it's super funny, but then again maybe it's just not that kind of joke.

I take a bite as Esme continues to narrate and find Linda immediately behind me, striving both to get a bite for herself and to hear what Esme's saying.

The slider tastes simply perfect. It is the platonic ideal of what a Reuben dreams it could become. The flavor of what I guess is pastrami lights up the whole bite but is also challenged by the acidity of the kraut and then mellowed by the creaminess of the cheese sauce—apparently the Mahon fondue and/or fancy sauce, whatever that is.

"So they take the beef heart," Esme continues, and I stop chewing right before I swallow. Heart?

"And they clean it up and then marinate it . . . or no . . . not marinate but rather brine it . . . cure it . . . for what was it, Chef? A week?"

"Yep," he says.

I don't know how he can even hear her over all of our bodies and the steady whoosh of the hoods above the cooking stations, but I get the early impression that he's apparently always listening. If it weren't for the shock of the fact that I've got a wad of a cow's heart in my mouth, I'd begin worrying that he too heard my stupid student loan/Springsteen comment.

"And then," Esme continues encyclopedically, "they smoke it for a day over hickory?"

"Applewood," the chef corrects as he tweezers the finishing touch on something that another cook picks up and begins to bring around to us. "Hickory would be too bitter."

"Smoke it over applewood," Esme says, "and then slice it super thin. And don't forget the little cornichon on top." She takes for herself the mini-pickle from the fanciest toothpick I've ever seen and pops it into her mouth. "Damned nice," she concludes.

She's right, and I swallow, figuring I can either succumb, acquiesce, or go away.

Whatever it's made of, it's simply mind-blowingly good.

I mutter something with gustatory approval, and Linda looks at me with her fierce blue eyes.

"So good," I say.

"Better be," she says but doesn't elaborate, and my heart shrinks a little.

Another couple of fascinating and thrilling dishes come out: a fried green tomato with pickled king crab, a tiny bowl of littleneck clams in a vichyssoise, and then caramelized cauliflower with a peanut and carrot puree that blows the lid off my culinary world. Cauliflower. Motherfucking cauliflower and I feel like I'll never taste the equal of any dish—so simple, so intensely flavorful, so memorable, so just perfect.

I check my watch and panic. Somehow nearly three hours have passed, and I need to leave to meet my son at the school's Running Club he's doing on Tuesdays and Thursdays this month. Ordinarily he takes

the bus home on these days, and, though it's far from ideal, he can at
least just let himself in the house with his key and watch some TV until
I get there, but today Running Club gets over at ten to four, and after
that he'd just be left alone outside the school. I have got to get there.
He's just seven, and while I might have been riding my bike to and from
school since I was a kindergartener, that's just not how kids are raised
these days. Not ours, anyway, and I am afraid if I'm any kind of late either he'll wither away into despair or they'll sick Children's Services on
us, or both.

I just have time for one last dish, and as it lands on the counter, Miko
announces that he thinks it's the pork.

"The jowl?" Hans says, with a slight air of inquisitiveness. In a grand
gesture for him, one of his eyebrows is arched in question in the general
direction of the kitchen.

"Yeah," Chef says, pausing now, letting his tweezers drop to the cutting board. "It's not the cheek or the neck," he says and grabs the un-
derside of his bearded chin. "It's that big fatty stuff on a hog. Practically
pure fucking fat. It's awesome." I like that as sophisticated as his food
is, he's still both vulgar and down-to-earth about it.

Esme is still taking the lead on describing the food, spinning the
plate around so that she and we all can see the dish from every side before she cuts it up with both her knife and her description.

I fork a bite, but I don't have time to listen to her elaboration and
instead just slide down to Dan to tell him I've got to run, but before
I can get the words out the flavor hits me. It tastes like a simple street
taco, but on methamphetamines. I expect the pork to give me a struggle, but rather there's just the slightest crunch of a crust and then just
this luscious pillowy pork essence inside. It doesn't taste like the fat
I'm afraid of but instead like pure flavor, and the flavor is familiar but
keeps coming in unanticipated waves. There's creamy cotija cheese, I'll
learn later, a smoky but smooth black bean puree, and then at the top
of the experience a hazelnut vinaigrette to lend just the idea of bright-
ness to the mix.

"It's kind of a joke," Chef says. "Just supposed to be my play on a
taco."

There's nothing funny about it. This thing has just totally kicked my
ass.

But then, as I tell Dan that I've got to go, Esme pipes up. "Maybe add some lime, Chef? Needs some acidity."

This seems like server suicide to me. Not only to make a suggestion about what is clearly an entirely thought-through dish but to do so in front of everybody—the whole front- *and* back-of-house staff. A beat goes by in which all chewing and knifing and forking stop. Even the ventilator hoods in the kitchen seem to stop their incessant whirring to see what will happen to Esme.

But Chef—apparently he's the chef—just shrugs and recrosses his arms so that the rooster tattoo is covered up by his other arm but the talons near his wrist are still visible. "Yeah," he says as though it was his idea in the first place. "Sure. Lime."

I go to grab my bag and my folder with all of Surly's trade secret recipes for both the beer and food, but it's not on the table. I had just left it where I sat, but now there's nothing there, nothing on the floor, and all of the folders at the tables where we were sitting before have other people's names on them.

Another dish has come out and everybody but me is diving wordlessly into it as I search in panic around the tables and the bar, trying to find my fucking folder.

Dan notices me and gives me a slight perturbed nod, the way two male rivals tend to do in urban settings where they're saying with just one gesture both "What's up?" and "You got a problem?"

"My folder," I say helplessly to Dan.

He smiles a toothless smile, but there's a wrinkle of irritation at the corners of his eyes. Not only am I skipping out of training earlier than anybody else, but I have somehow managed to lose the folder of proprietary information without even leaving the room.

"That's okay," Dan says, but it sounds more like "strike one" in my head. "I'll make you another packet for tomorrow."

I can only apologize so much, I know, without appearing to be a complete putz, so I take off, running to my faraway car when I recognize that not only did I leave early, fail to taste all of the dishes, and lose my folder so I won't be able to study tonight. I also forgot to clock out.

Chapter Three

HEAVEN CITY

IN THE WORLD WHERE I WAS ONCE A CHILD AND NOT YET A man or a husband or a father—only a son—my mom and I, after nearly an hour of bumbling around the Wisconsin countryside, found ourselves just outside Mukwonago at a kind of ramshackle resort restaurant rumored to have been a favorite of Al Capone back in the day. Heaven City. As its name might hint, it began as something of a religious commune, and the moniker stuck long after the monks with their hymnals were swapped out for gangsters with their tommy guns. Built on the rolling pastureland of greater Waukesha County, the restaurant is a blocky building with deep eaves that looks like a square, two-tiered layer cake. It almost evokes the prairie poetic architecture of Wisconsin's own Frank Lloyd Wright, but it's a bit too old and a bit too boxy. Still, the building is firmly placed on a grassy plot on Phantom Lake, as beautiful as anything gets in Wisconsin.

Technically, Mukwonago is a village. I only knew it as the place a handful of my high school friends were from—the place where other classmates would joke that all it was known/named for was the fact that not even the dirt wanted to stay there. Muck-wanna-go. Turns out it means "place of the bear" in the Potawatomi language, but I knew nothing of that at the time.

All I knew was that my mom wanted to take me out to a nice dinner, just me and her. I'm pretty sure it was on the eve of her second divorce, though I don't remember the occasion now. And I'm pretty sure I drove, because rather than the dark drive being boring as was my memory of long, dark, winter drives in Waukesha County as a kid, now it was terrifying because I was behind the wheel and had never driven this far from home, never mind the fact that my mom was ostensibly navigating.

It didn't matter much. Waukesha County, outside of its biggest namesake city of Waukesha, is actually pretty rural once you're off I-94 or I-43, the two interstates that almost define its north and south boundaries. We were heading southwest on National Avenue, the street both my mom and I worked on in New Berlin, some fifteen miles away. But National out here resembled not the domesticated scene of our second-tier suburb but rather just random, wide-open, black-skied rural America anywhere at night. Addresses out here are more like co-ordinates than street numbers, and I'm sure if you have a sextant and can navigate by the stars, an address like S91 W27850 would be a piece of cake to find. For us, alas, it was not.

So, after passing the place and missing it by miles before turning back and missing it again, we finally found Heaven City.

Nothing marked this destination restaurant—a phrase that didn't exist then—but it certainly was one. It just had a little white sign no bigger than a cafeteria tray with script black letters on it and a winding drive leading off the road and up a hill to the expanse of an unruly, un-lined crushed-stone parking lot surrounding the restaurant. Though it did have an air strip to attract the well-heeled crowd from Milwaukee and Chicago for the weekend.

Getting out of our car, hearing the crunch of the gravel under our feet, letting our eyes adjust to the deep darkness of rural outer-space-ness with its lack of light pollution and also just plain lack of lights, we took in the solitude, darkness, and quiet of the space that seemed like the kind of place one goes to work as an undertaker or a serial killer. Its stillness was too absolute. Though we could only corrupt it by our pres-ence, and it felt like a violation we would likely not be able to undo, still we persisted, arm in arm, and headed up to the door.

My mother was wearing (I hope, though truthfully I don't remem-ber) her rabbit fur coat. It was a motley thing to look at, stitched to-gether with the pelts of dozens of bunnies, but to feel it was sublime. Politically I'm about as far from pro-fur as it gets—and, for what it's worth, I only eat meat these days maybe once or twice a month even though I live in the Midwest and am married to the daughter of a Nebraskan cattle rancher. But when I was little, say, six or seven, when my memories are little flurries of sensory overload rather than orderly narratives with beginnings, middles, and ends, I remember those little bright spots of time right before my mom and dad would be about to head out to a nice dinner. I knew it wasn't just a night out to their favorite gin joint, the Black Kettle, from whence they'd come reeking of ashes and onion rings, but someplace fancy—a chop house or a proper supper club, which was pretty much the whole gamut of their fine dining options in the seventies and into the eighties in Wisconsin. My dad would noose himself up into a tie and a sport coat with lapels wide enough to land a plane on, and my mom would wear a dress—something she absolutely never did except to weddings and funeral. And just before they'd head out the door, she'd reach deep into the front closet, back beyond our usual thick winter coats stuffed with the feathers of geese or, more likely, chickens, back beyond their spring-weight windbreakers—a name that would only become funny to me later—back even beyond my bright royal blue satin—yes, satin—Milwaukee Brewers' jacket that I loved but was usually too embarrassed to wear since not only was it satin but it also said something about being a member of "Harvey's Wallbangers," so named for coach Harvey Kuenn and the Brewers like Cecil Cooper, Ben Ogilvie, and Stormin' Gorman Thomas and their hard-swinging (but rarely home-run-hitting) ways—and there, in the very back beyond of the closet, my mom would pull out her rabbit fur coat.

Hearing the squeak of the door as it was leveraged to get that deep into the closet, I would fly from the cold embrace of the basement and my Atari and dive into my mother's warmth as my dad helped her from behind into her coat. The fur on the outside was in turns sweetly smooth and delicately furry—far more downy than the coat of Sammy, our recalcitrant spaniel—but the patches jaggedly bordered each other and would remind me decades later of the stonily segregated pastureland of the Aran Islands, where my wife and I would spend part of our honeymoon. Whatever tailor sutured this suit of fur together was not precious about his trade, and, accordingly, the coat was not much to look at—it could have been made by an eighteenth-century fur trapper with a bone needle and horsehair thread or it could've been made in a Mumbai sweatshop or it could've been made by Ed Gein or any number of Wisconsin's many serial killers. It didn't matter.

The inside, though—the inside of the jacket was the real secret. Strange and schizophrenic as the outside was, the inside of the jacket was made of the same satin as my Brewers' coat, and just before my parents would leave, my mother would envelop me inside that coat, and I would be again forever ensconced in the milky sweet smoothness, and everything, for that solitary forever moment, would be okay again.

It all seems pretty obvious and Oedipal and so forth now, but there's no telling that to a seven-year-old. My mom was the most beautiful woman in the world, and when, in the Freudian dark dream of that night in the Heaven City parking lot, when my mom took my arm in hers with her rabbit fur coat on, I was, once more, once more, her baby.

As I would soon enough never be again.

In high school, I got a job as soon as I was old enough to work, at a place called the Torque Center. It was a bike, skateboard, and motocross store just a few blocks down National Avenue from my mom's flower shop. I liked the discount on bike and skateboard parts, and a few of the staff were among my favorite people ever, including sweet Teresa, who was so short she had to wear heels to outdo a yardstick; Mike McNamee, a kind and super talented freestyle and vert rider who could do unnatural but beautifully acrobatic things with bikes; and, of course, Arlo and Duane, the southeastern Wisconsin unironic answer to Chicago's Wayne and Garth. They were all in their early twenties and were generous and protective of me and did their best to shield me from the growling, finger-snapping assistant manager, Todd, and the rest of his hateful motocross crew. It was a great first job, but it didn't take me long to learn that there's very little magic in the world of retail.

From a relatively early age, I could tell that there was something different about restaurants. Sure, they were places of business and were out to make a profit by charging more for what they sell than what it cost them to buy and prepare it, but somewhere along the way, something almost like transubstantiation takes place.

Perhaps it has a lot to do with the fact that as a kid, all the adults in my life worked full-time and not a one of them—not a parent or an aunt or even a friend's dad—not a single one of them could or would cook. My childhood plates were populated with Van de Kamp's frozen clams with the consistency of old fan belts, geometrically regular obelisks of fish, and chicken nuggets that invariably came, like the tract houses in my subdivision, in one of four shapes. There was the puck, the lazy diamond, the chubby hourglass, and, everybody's favorite because it lacked the suspicious, unnatural regularity of the others: the ear. They all tasted the same, like sawdust reconstituted with a little meat glue and chicken fat. Every now and then somebody would try to grill burgers by making huge fists of ground beef and throwing them on an unsuspecting grill, but they always ended up charred on the outside while remaining horrifyingly raw in the middle, tasting of nothing except the petrochemical nightmare of lighter fluid since nobody in my sad suburb could figure out another way to light charcoal on fire.

Hence, even if they were mediocre by most gustatory standards, even relatively shitty restaurants were hallowed in my youth for being able not just to give me something to eat but to actually pique my attention, my imagination, and, of course, my appetite.

Here, then, is a brief litany of the restaurants of my youth.

There was, as far back as I can remember, The Chanticleer. A poetic, cheeky name for a fried chicken joint in the middle of rural Illinois, a few cornfields away from where my grandparents lived. Its parking lot was bordered on three sides by soybeans, but despite its lack of an appealing location—there was literally not another structure in sight, and this was central Illinois, so you could see for miles in any direction—it was always packed. Before corporate restaurants like Fridays and Chili's existed, The Chanticleer made it fashionable to nail rusty farm implements, expired license plates, and oily car parts to the wall and call it decor. I now live in a city where a similar restaurant, Revival, simply cooks the same stuff—southern-inspired soulful food like greens, hoppin' John, and fried chicken—but has convinced the culinary elite that this is haute cuisine. Revival is helmed by a James Beard award–nominated chef, but it was his chicken joint, not his fine dining restaurant, that was recently named restaurant of the year by the local press. Ask anybody in rural America why. It's not fancy and it's not pretty and it's not supposed to be served on white linen tablecloths. It's fried goddamned chicken. I doubt I'll ever have a better meal than those at The Chanticleer.

Then there was my grandparents' similarly rural and rustic but deceptively categorized Pekin Country Club. Yes, there was a golf course and a swimming pool and there were chefs with white paper toques and if you were a member you could pay for your bill by just signing your name, but nobody who had been to a city the size of Peoria or bigger was fooled. It was a mediocre restaurant that usually didn't try any harder than to serve a rectangle of meat with a pile of starch with some reddish brown sauce from a can if it was dinner, a muffin, eggs, and a nuclear yellow sauce from a can if it was brunch. Except, one random day in July—I imagine it was July, because when else but in the crucible of an Illinois summer could someone come up with something as miraculous and unexpected as this: cheeseburger chowder. Sure, it was probably just Velveeta with browned ground beef in it, topped off with a handful of croutons, but by damn it tasted like a cheeseburger.

No, it's not molecular gastronomy. As far as I'm concerned, it's better. Better because it's not trying to fool anybody like so many *Top Chef*-y cooks are trying to by deconstructing rustic dishes or overly complicating beautifully simple things (*and here we have a play on a po'boy with the Santa Barbara spot prawns extruded into a "roll" and the bread gelatinized into "fried shrimp"*). Nope, that is never going to fly in Pekin, Illinois. It's a cheeseburger, but in soup form. Nothing to think about, just something to eat while you wait for your turn on the back nine.

And then there was Kopp's Frozen Custard and Burgers in Brookfield, Wisconsin. Frozen custard, if you're not familiar, is basically ice cream all hopped up on extra fat and eggs so that it's not hard or icy but rather rich and creamy and guaranteed to pop your belt a notch or two. What I loved about Kopp's was that it's one of the first restaurants I ever encountered that decided they were going to serve two things. Frozen custard and burgers. Their frozen custard was great, but it was simply on par with LeDuc's in Wales or Lixx's on the East Side of Milwaukee or Leon's famous drive-in in West Allis or any number of Wisconsin establishments. But their burgers? Kopp's burgers were about the size of a school bus steering wheel. They weren't thick or particularly interestingly seasoned, but they were sized unlike anybody else's. Which meant, even to me as a kid, that they made them there. And it also meant that they made the buns there—or at the very least had them special made just for them. All of which is a long way to say it opened my eyes to the fact that restaurants didn't have to be simply openers of cans and boxes and reheaters of already cooked food. They could decide they were going to do it *their* way even if they were the only ones. Maybe this falls somewhat short of Ezra Pound's modernist imperative to "make it new!" but I think not. Because not only was it new to me to get a burger at Kopp's, but it was also special. (It was also way too much food, but this is Wisconsin we're talking about.)

And then in Milwaukee there was The Chancery. It was something of a local chain restaurant—I think at one point there were three of them in southeastern Wisconsin—but it was still a decidedly local restaurant. Unlimited fish fry on Friday—and not just during Lent either— cheese curds of various iterations (all fried, of course), and then this dish that was an absolute revelation to me. I remember ordering it at their now-defunct East Side location right next to the Downer Theatre and Harry W. Schwartz Bookshop. It was something I'd never heard of, never mind knew how to pronounce. Andouille Pasta. I don't remember if that was the actual title—I'm afraid it was something like "Hey Curly, Where's Mo?"—but I remember being lovingly tripped up in the mess of vowels that were somehow vaguely French but I knew instinctively to be either Cajun or Creole. It was the first time I had pasta that was spicy and creamy, and I had likely just come off yet another bruising trumpet lesson at the hands of some music professor from the University of Wisconsin–Milwaukee. I'm still not sure how I got paired up with him as a freshman in high school, but suffice it to say I didn't do well in his eyes, and I almost never played trumpet again after I barely endured a few taunting lessons with him. Nonetheless, I loved the trips to the East Side with my mom, not because of what I learned but because we started to eat regularly at The Chancery, and that andouille sausage dish always made me feel better.

And for what it's worth, since no one in my family ever cooked meals, restaurants would become the linchpin of our family time, and no single restaurant became that to me and my parents more so than The Chancery. Shortly after our meal at Heaven City—and shortly after their split-up was official—my dad was living at his sister's house, recovering from open-heart surgery (and divorce), and once he was well enough, we'd go out to the South Side Chancery by General Mitchell International Airport. One week I'd have cut and puffy lips from another crushing trumpet lesson and my mom would take me to the East Side Chancery and call me Sweetheart and try to put on a brave face about having hopefully made the right decision, and I'd scarf the free popcorn and order that andouille dish. And then, the next week, I'd go see my dad and help him prick his finger to test his blood sugar and then we'd go to the South Side Chancery and he'd have to settle on an iced tea and a salad instead of a burger and three martinis. He'd call me Tiger and try to talk about cross-country, or preferably any other sport on earth, but I didn't know anything about any sport other than running (at which I was likewise pretty dim and untalented), and I'd scarf the popcorn and order that andouille dish. A direct, mutual kind of love never seemed to be quite available to us all, but The Chancery was our only shabby grail.

And then, flashing forward quite some time, there was the Water Street Brewery. This was the early nineties. Brewpubs were just barely a thing, especially in the Midwest. I'd never seen a place that made both food and their own beer—and also made strides in both areas to make them unique from all the rest of the typical offerings. I don't recall most of their menu, but I remember it was printed weekly in the format of a newspaper, which was then and remains still very cool. And, of course, I remember the stainless steel tanks and copper kettles where they made at least a dozen different beers right on-site, but what I remember most about Water Street were their pretzels and mustard. Not exactly a remarkable dish on the surface of it, but they made both of them in-house. And to have fresh-baked, still-hot pretzels in their wonky irregular sizes—because, I realized, an actual person made them, not just some machine—and their non-school-bus-yellow spicy brown mustard full of seeds, zingy with vinegar, was a revelation to me. At first it seemed preposterous to pay six or seven bucks for something we could get at the Piggly Wiggly for less than a fifth of that, but once they arrived at the table, it was clear that there was no replacing what we were getting. Those pretzels were epiphanically special.

Dining at the Water Street Brewery was not only a turning point in how I felt about scratch cooking, it was also really the first time I thought *I want to work here.*

I've come to love many other restaurants over the years, and most of them I mention by way of eulogy rather than review since they're almost all just memories now. There was the exuberant and eclectic Ziggy's gyro place on Wisconsin Avenue while I was in college in Milwaukee, where Ziggy smoked an actual pipe while he sliced the charred meat directly into the catcher's mitt of the pita. There were the ubiquitous corner delis in Boston where I began grad school and practically lived on their chicken parmigiana "grinders" (sub sandwich to everybody else). In Brookline, there was Matt's Pub, which I liked if only because I'd never been in a single establishment before that shared my name. But it was at least somewhat authentically Irish, and their fish and chips were as much of a new dawn in my life as was their homemade ketchup. There was the asparagus risotto at Michelangelo's in Salt Lake City, where there was also sushi I enjoyed with my beloved mentor, Melanie Rae Thon, at a former church called Ichiban and where there was also more Cajun cooking at The Bayou on South Main Street with their hoppin' John (their secret being, I figured out after probably a hundred attempts to reproduce the dish at home, butter, crispy bacon, and fresh chopped green onions all garnished just before serving).

And there would be the restaurants I'd more revere than love, not because of what we ate but because of why we ate there, such as the small marble table in the bar at W. A. Frost's in Saint Paul where Jenae and I split a very good burger and some nicely crisped calamari after we closed on our house. There was the universally terrible Champp's (*sic*), where we ate with my mom and her new husband when Jenae realized the reason she couldn't finish her Bloody Mary was because she was pregnant. And then there was another local chain restaurant, Famous Dave's, where I picked up brisket and chopped smoked pork to take back to Abbott Northwestern Hospital after the birth of our son, and, the next day, after some restaurant-savvy nurses helped give us a better lay of the carryout land, True Thai on Franklin Avenue for some gringo (but brilliant) cream cheese wontons and far-as-I-know quite authentic green curry.

And then there was the little Fort Atkinson, Wisconsin, supper club where my mom and I met—her driving halfway from Waukesha, me driving halfway from Madison—to celebrate the fact that the grapefruit-sized mass they found on her abdomen was benign. All I remember was that the table between us had one of those glowing orange orb candles and that I wore a bow tie because she always liked them and that we both probably had too much to drink in celebration of her all too briefly clean bill of health.

Having worked at nearly a dozen restaurants in my life, I have a pretty good idea as to what any given kitchen looks like, as well as its walk-ins and dish pits and dry storage and loading docks and server stations and everything else, from both the perspective of the diners and that of an employee. They are almost never as pretty or as clean and tidy (and free of, say, standing water and rat traps) as you'd hope for. Which to me makes the theater of dining out all the more compelling. That we not only mostly don't get sick but, unless there's something wrong with us, usually end up having a pretty good time.

To be sure, lots of people go to restaurants in order to not have a good time. They don't know it, but they do. They expect to be let down, to have to wait more than four minutes for their stupid Brandy Alexander, to have to endure yet another server who doesn't know what every spice in the Togarashi blend is or who is so ignorant as to know neither the age nor the gender nor the breed of the grass-fed organic beef they're serving, to not be able to simply enjoy a meal because they don't want to order what the restaurant is preparing and advertising and actually selling but rather want to order whatever they want because that's how they go through life.

But their anger and disappointment is, of course, their own reward.

I have high standards and don't expect to always be blown away, given how often I've eaten out and how many places I've worked, but I have found that working in restaurants gives those who do it a ready-to-go sense of empathy and compassion toward fellow industry workers. If your world is falling apart as a server, it usually only takes a glance around to realize there's probably a perfect storm of shit swirling around you and it's not your fault.

Everybody who has ever worked in the service industry has had apocalyptically bad days, and they follow not just our waking selves but also forever into the lives of our dreams.

You have a brand new host who will fill up your eight-table section all at once before moving to the next server's section because why not? Your bartender is busy flirting with that sweet little something at the far end of the bar and your drink tickets just keep lining up like the unanswered prayers of the destitute. Your manager is hiding in the office because he's got a double-barreled hangover and can't bear even the sound of the printer, never mind the din of the dining room or the clamor of the kitchen. You have nobody to expedite or run your food, so whenever it finally does make it to your table, it's wrong or cold or just sad. You only have one pen and it's out of ink. Your printer just ran out of paper. Everybody at that ten-top who has been playing musical chairs all night wants separate checks—no, scratch that, only some want separate checks. Others would like to pay for that one guy's drink and that one appetizer but they didn't get the chardonnay but rather the sauvignon blanc (which are thirty cents different in price), and could you put the food on this card and give us an itemized receipt for this and put the booze on this other card? And, oh, by the way, your sweet but handsomely oblivious host has just told you she sat a table of eight for you in somebody else's section. And also they're all VIPs on Yelp.

Ah, restaurant dreams. I still have them. I've never once had a bad dream about any other job—not retail, not the breakage room at a beer distributorship where I briefly worked, not even the awful administrative jobs or the ESL teaching job at Berlitz—not any bad dreams about teaching ever. But the restaurant dreams I've had all my life and still do.

The fact of the matter is—and don't we all know this?—most restaurants are super high-stress environments that put the souls of the poor, the undocumented, the undereducated, the criminally offending, the substance abusive, the single parenting, the night schooling, the aspiring writing/acting/singing through the most brutal of workplace woodchippers. Because you know who mostly isn't working in a restaurant? The entitled, the silver spooners, the trust funders, the lazy, the semiprofessional skiers and/or snowboarders.

Actually, there are a lot of semiprofessional skiers and/or snowboarders in the service industry, because they too have to work their asses off at night so they can ski all day. And, actually, there are also very often super talented, hypereducated, and tremendously driven and career-oriented folks in food service. Maybe even some trust funders. But not many. Because that's also the beauty of most restaurants: if you can do the job, they don't care who you are.

And that's the great leveling blade of the service industry. It's hard. There's no place to hide. If you suck, everybody will see it and blame you and will perhaps even write nasty passive-aggressive notes on their credit card slips and/or actually aggressive (but pathetically anonymous) reviews online. If you work in a restaurant, you will either sink to the bottom and be fired or just not scheduled or you will be forced to work the tedious opening shifts during the week where you'll make—maybe—your bus fare back home in tips until you can start picking up a brunch here or there, and then, finally, somebody else has to take care of their sick kid or they get stuck in an avalanche or whatever, but you—you answer both the proverbial call and your manager's panicked text, and you say Yes, I can work tonight.

And pretty much like that, you're in. You are a part of the troupe, the company, the brood, the brigade. And with every shift, you're going to have the opportunity to prove to everybody that they were right to trust you to serve alongside them. And also with every shift, you've got the opportunity to decide that maybe you should go back to folding T-shirts or sweatpants at the mall at whatever retail abomination is currently responsible for putting printed words on young people's butts.

All of which is a very long way of saying it's hard work and it's really not only no surprise when dining out is less than stellar but rather pretty much a freaking miracle when things go right. When you get the drink you ordered. When your appetizer comes, like you had hoped, before the other food, as, arguably, appetizers should. When your very reasonable (but, it's true, somewhat sweetly provincial) request that your steak be cooked to medium-well/well comes to fruition. When, weeks in advance, you let the host know that it is your darling bunny's birthday, but then, upon arrival, you didn't say anything else, and hence you suffer through the whole meal with a vague sense of doom that they won't have remembered, when, Lo! They remembered! And that special sugary frosted thing with a tiny flame atop a tiny candle is coming for no table other than your own.

Because, ultimately, that's the goal if we're being our best selves: to make you feel, simply, special.

But long is the way and narrow is the path—and also there are more people on the books than we actually have seats for this evening, so good luck with that. So really, it's not unusual that restaurants are often not great. What's unusual if you've ever worked in one is that they're not always utterly catastrophic.

Which is why the good ones linger with us, as profound and memorable as gifts given to us by our own children—special not simply because our kids made them but because they actually put some thought and effort and love into them.

As was the case that one night at Heaven City.

It was dim but not dark. There was light, but it was a secretive light. It emanated from sconces and recessed fixtures. Light came from focused beams overhead. Light whispered out from shaded candles on every table. Light there was dramatic, purposeful. It allowed us to see what was important but not to fret over every detail.

The host greeted us and checked us off on her great leather-bound ledger and took us through the bar and into a kind of atrium—it had to be where the second story met the first—but instead of two levels, this was one room with twice the amplitude, and beside our table were flourishing palm trees, the likes of which I'd only ever seen in Florida or those domed botanical gardens in downtown Milwaukee on grade school field trips. This was luxurious and exotic, but at the same time it wasn't stuffy or pretentious. It was nice, but not in an exclusionary way. I was glad to be wearing a shirt and tie, but I didn't feel like an outcast for not wearing a sport coat.

Not far from our table there was a jazz trio. A piano, a tenderly restrained drummer on just a snare and a high hat, and a saxophone. It was schmaltzy but sweet, and irony wasn't yet a concept I knew anything about, so there was no cloudy self-consciousness to fog the night. It was nice. It was very nice.

First, there was surely a vodka tonic for her and a Sprecher root beer for me. Then, likely as not, there was something Midwesternly typical such as shrimp cocktail but scaled up, and so it might've been served in a martini glass with the tails hanging off the sides like garish eighties earrings and with spicy, horseradish-forward, homemade cocktail sauce. And then, in accordance with the restaurant's prescient mission—it wouldn't become trendy and all the rage until years later after somebody like Alice Waters, Charlie Trotter, or Thomas Keller could claim it—there was local food and game, prepared and presented in what we now call a farm-to-table way. There was local pheasant served simply with the kinds of nuts and berries it, too, liked to eat. There was roasted pork tenderloin with a compote of apples from a nearby orchard. And they were good. They were very, very good—like nothing I'd ever had before, at home or in a restaurant, anywhere. But they were no preparation for what would come last.

Dessert, strangely, was something we had to order early, right after our appetizers and before our entrées even appeared. There was no Cooking Channel then. No *Top Chef.* No *Hell's Kitchen.* No Emeril, even. The only chefs I'd ever heard of at that point were Julia Child—whom no one in my family had any awareness of—and Justin Wilson, the Louisiana chef I liked because of his entertaining Cajun patois and because he would say stuff like, "Now you put a little dat wine in the pot, den you put a little dat wine in you," but other than that one dish at The Chancery, I had no concept of French or Louisiana cooking. Suffice it to say, I had no idea what a soufflé was or why you had to order it so far in advance or why it was something of a big deal.

Though there were no sparklers or candles or throngs of servers belting out "Happy Birthday" while clapping like those wind-up cymballed monkeys, I felt the eyes of all the other diners track our server as he made his way through the dining room to our little palm-shaded table with our soufflé. A second server followed behind our own, and he was carrying a little silver pitcher that he held aloft with all of the ceremony and seriousness of an altar boy, but all of his attention was on the first server, who held the plate that held the little white ramekin with the poof of our soufflé defiantly parading above the rim. He held the plate as gingerly as though it were a bird that had miraculously landed on his hand. Gently, gently, our server set the plate down, the ramekin sliding a little, almost off the paper doily, but then correcting, and the plate was on the table and the ramekin was still on the doily on the plate. And then the second server swooped in and announced, "Your Grand Marnier soufflé with Frangelico crème anglaise," and he took a spoon and popped the top of the soufflé in the center and quickly poured the anglaise into the hole and then around the top and then he whisked himself away.

I don't believe either my mom or I had ever seen anything presented quite like this, and I was a little skeptical as to what all the drama was about, but the urgency of the situation was upon us and neither of us hesitated and we both dove in with spoons, and, tasting, we made eye contact across the candlelight and I could feel both of our consciousnesses expand.

There was orange. There was almond. There was cream, but it wasn't milk—it was silky. Sweet. Supersweet, but somehow not cloying. There was hot. Heat. It was boozy, and the burn of the alcohol lingered on my lips and on the back of my throat, but so did the citrus and the nuttiness of the cream. Mostly, however, there was air. The soufflé seemed to be made almost entirely of air. Clearly other ingredients were involved, but air, hope, and prayer seemed to be its main constituents, and they came together in as holy a fashion as any trinity I've ever known.

And then it was gone and then we had to leave and then we were back out in the blackness of the unlit Mukwonago night, and soon my mom would be divorced again and we wouldn't live in our house anymore and there would be a crappy apartment and the food there would only be heated, not prepared or cooked in any meaningful way, but that would all be okay. Eventually it would all be okay. That night in Heaven City couldn't last, of course, nor could it really nourish us for any meaningful period of time beyond, say, a dozen hours at best, and yet it lived on for decades in both her and my memory as a perfect evening, crystalized and suspended in amber.

Which is what restaurants can do. They cook so you don't have to. They make you feel at home, even though you aren't—maybe even more at home than you feel at home because here your mom or uncle or grandma or whoever isn't in the kitchen. The kids aren't in the basement. Your dad isn't glued to the game on the tube. You are all at the table. There's a word for this, of course: communion.

Restaurants, per se, didn't quite exist yet, but when it came time for all the disciples, even Judas, to eat, they decided to find someplace nice that could fit them all rather than just head to one of their houses. Not even Jesus wanted to eat at home for his last meal. He wanted to go out.

* * *

And now, a little more than two decades after that night at Heaven City, my mom has breast cancer. Something specifically called a "triple negative" cancer that's not responsive to hormone therapy or surgery or radiation, and so chemotherapy is where we start. Two weeks before Christmas, her main concern was that my in-laws were going to bring their RV to town and park it in my driveway. One week before Christmas, her main concern was that she found a lump. Three days before Christmas, we met her surgeon. She had the results.

On the way home, we stop for drinks at The Lexington, Saint Paul's jazz-era supper club. We manage to find, despite a loud and bustling bar, a quiet room behind it with two leather club chairs next to a fireplace. We talk about my son's newfound love of playing goalie in hockey. We talk about how to best pick up dog poop when you're wearing mittens and it's below zero. We talk about cancer.

The feel of The Lexington—the dark wood beams, the warmth of the lighting, the way my mom's vodka tonic looks special, somehow like a potion in its heavy crystal glass—it reminds me of Heaven City.

I ask her if she remembers that meal, her rabbit fur coat, and that soufflé.

She is about to take a sip of her drink but stops, the ice gentle against the glass in the air.

"Of course," she says. "Of course I do."

There's so much we don't know about what is going to happen, but for right now, we've got each other, these chairs, this fire, and our drinks. I raise my glass and touch it to hers.

Chapter Four

SCHADENFREUDE

AFTER I LEAVE SURLY, WHILE WAITING AT A TRAFFIC SIGNAL on the way to my son's school, I take a chance and send my neighbor Heather a text asking her to tell Emory to just hang out after Running Club by the jungle gym until I get there. That I'm on my way and won't be more than a minute or two late—I hope—if I can find a parking spot anywhere near his school.

I'm struggling to turn left out of the industrial park in which Surly is situated; I'm stuck behind at least two semis that are also trying to turn left onto University Avenue. I'm probably only two miles as the crow flies from my boy's school, but it might take me one or two rotations of the light just to get properly on the way.

Just as the truck in front of me swings into the turn I get a reply from Heather.

"Sorry," she writes. "Have my seminar tonight. No RC for me."

Fuck. So now not only can she not help me out, but I've also made it abundantly clear to her—she is also a colleague at my school—that I'm a shitty, negligent parent.

After some very specious driving, however, I make it just in time to see the last runners straggling toward the finish. Running Club, it turns out, is more like sprint-then-run-then-jog-then-walk-then-just-putz-around-the-park club. My boy is so far back in what's left of the pack that as far as he's concerned, I got here early.

It's little consolation, though. Something about him just doesn't look right. It's a sticky, warm spring day, and maybe, I think, he's just a little overheated or dehydrated. Usually when I pick him up from Running Club I take him for ice cream after, but I ask him today if he'd rather just go home.

"Yeah," he says from his booster seat in back. "I just want to hang out with my puppy. I'm tired."

It's been a long school year for him with a teacher he never really hit his stride with, and it just seems like he's at his limit. Running Club, while great in concept, is a little bit too much too late in the year, I think.

When we get home, he piles up on the couch with Frankie and I go to the kitchen to see what kind of ice cream treats we have in the freezer, and then I hear what sounds like a bucket of water being poured on the floor. I shut the freezer and run to the living room, prepared, I'm afraid, to be highly irritated with Emory for whatever he just spilled—wondering simultaneously what it could have been, considering I haven't got him anything to drink yet—when I see vomit spewing from his mouth right onto the living room carpet.

I swoop in and pick him up and carry him to the bathroom, pathetically concerned both for him and for our upholstery should he start spraying vomit on the hand-me-down furniture too. But by the time I get him in the bathroom, he seems to have at least momentarily exhausted himself.

Unfortunately for him and us too, this continues well into the evening. Jenae is, as she herself describes, a sympathy puker, and I worry that once she sees Emory hurl, she'll give it up too. I struggle to think of what it could have been—we all had the same pasta for dinner last night and Emory's lunch today was the same Wowbutter (he's allergic to peanuts) and jelly sandwich and applesauce I sent him with yesterday, and he swears he didn't have anything else out of the ordinary, so we figure as long as he doesn't get a bad fever it's probably just a nasty stomach bug, the likes of which are about as common at his age as cooties.

If it weren't for the fact that I've already had to make an exception for myself to leave training early today, maybe I wouldn't feel so anxious about tomorrow, but I've got more mandatory training at both Surly and the other place.

"What the hell are we going to do?" I ask Jenae once Emory is resting peacefully in his room, a big white bowl by his side just in case.

She levels a flat stare at me that means what I'm asking is actually more like, "What are you going to do about your less important job so

I don't have to miss any of my more important commitments tomorrow?"

We both know without saying that school policy is clear; you can't send a kid to school after a fever or puking until he's twenty-four hours vomit/fever free.

"Well," Jenae says, "nothing for me to do but cancel my afternoon meetings I guess." She takes it on the chin as she always does when my stuff trumps hers. It hasn't happened often up until now, as most of my obligations over the past several years have been easily foreseeable ones such as attending a literary event or a conference or whatnot. My class schedule we usually know months in advance, so there are normally few surprises on my end. But now I've got two mandatory meetings tomorrow, plus another two on Thursday, as well as Friday, and Saturday, and even Sunday too, which is, looking at the calendar instead of just my training schedule, Mother's Day.

Shit—I forgot. I have nothing to give her.

None of us exactly gets a ton of sleep, but Jenae is up and out the door before either Emory or I are awake, and she returns home around 11:30 so I can scoot over to Surly. We barely have time to exchange formalities and I give her the update that the boy is hanging in there and has been, knock on wood, puke free all morning.

I'm riddled with worry about him as well as that it's day three of Surly training and I have yet to really lay eyes on the menu, never mind study it, since I lost my folder with all of their top secret files. But when I arrive, right where I sat the day before is a folder with my name on it. Somebody almost certainly grabbed mine by mistake when we were tasting the food and it's a huge relief.

When Dan arrives a minute later toting a new folder, I show him my old one as though it's proof that I'm a good boy.

"What the hell?" he says. He seems amused and irritated at once.

I tell him I don't know, and he flops the folder down on a table behind him and our training gets underway.

Today we're working in pairs—one new person with one of the folks who already has been working downstairs. I've been worried that there's going to be a constant us/them divide. I wouldn't blame them if there is. They've worked with one another for the better part of six months getting a new, big, busy restaurant up off the ground. Sure, we're doing the same thing on a smaller but higher-end scale up here, but nevertheless they were a crew before we've been a crew.

I imagine there was some thinking that it might even be better if the whole staff of our new restaurant came from downstairs—or if they were all new—but either way probably presented too great a poaching problem from their existing staff or the risk that an entire band of interlopers would descend on the brewery to serve at the high-end place, thereby kind of implying that nobody downstairs was good enough to work upstairs.

It'll be an issue, for sure, I just hope not a big or persistent one.

Fortunately, Dan pairs me with my (and Jenae's) favorite, Cate. There's something gregariously maternal and affectionate about her that I'm magnetically attracted to—that everybody among the staff is. Still, her eyes brighten when Dan announces that she's paired with me, though I have no doubt she would be as enthusiastic with any pairing. She's just a supersmart, supersweet person. She also clearly knows her shit backward and forward and could probably run the entire restaurant by herself if need be and still do it with warmth and genuineness. She also likes to call everybody "squirrelfriend," so there's that.

While Dan gives us the parameters of our task, Cate gives my arm a sweet, reassuring squeeze, and for the first time since the training at Surly has started, I finally feel like I might have a place here.

Thus far, it hasn't been a matter of wanting to belong. I just haven't. Needing to leave early yesterday and misplacing my folder—never mind not clocking out—really made me feel like a fuckup and an outcast. The first day was fine but almost entirely impersonal. I think, to be honest, it would have been better if Nichole ran the day's training instead of getting taken over by Gary. He's super knowledgeable, but so officious it's difficult to feel human around him.

But sitting next to Cate and knowing that for the rest of training today she's going to be my guardian angel makes me feel relieved down to my bones. It's as though up until now I've been wearing one of those beanies with the propeller on top, a KICK ME sign taped to my back.

I zone back in and catch the tail end of what Dan is saying about the rest of the week and how the main parts of the menu training are going to be on Friday and Saturday, and I fight the urge to check my phone to see if Jenae has any updates on how our boy is feeling, but a glance around the room tells me that I'm now the only one who hasn't made his own flashcards for the menu, and I sincerely wonder if I'll ever be able to catch up and get my head fully in the game.

Fortunately, as we begin to talk about the point-of-sale, or POS, system, much rings a bell. It's a stalwart system called Micros, and I'm pretty sure it's the same one I used at my last restaurant job in Salt Lake City.

"If you've used Micros in the past ten years," Dan says with an air of bemusement, "the good and bad news is that it pretty much hasn't changed. It's not pretty, but it works."

One by one, Dan tells us new folks our server numbers, and when he gets to me, he looks up and smiles.

"Five hundred," he says. "That's a keeper."

I feel a little surge of joy at this news. It probably sounds ridiculous, but this is the number I'll have to punch in the system every time I use it—within any given shift I'll probably need to enter this a hundred times. Within a couple of weeks, it'll practically be burned into my muscle memory such that I won't even have to look at the screen to type it in. Five hundred. It feels auspicious and I can't wait to tell other people.

Though I'm no numerologist, I've loved the number 5 forever. I remember it from the five-pointed star on Sir Gawain's shield—it was the topic of the first essay in college I got an A on and encouraged me on the path to become an English major—and I always loved the William Carlos Williams's poem "The Great Figure" about the "figure five / in gold / on a red / firetruck" and remember staring at the painting it inspired by Charles Demuth, a replica of which hung in the room at Ohio State where my eyes would invariably be drawn whenever a visiting writer went on a bit too long or veered inconsolably into concrete poetry or the like.

We set off in groups of six to tackle the POS training, and in our group, Rikki, a veteran server from downstairs, maybe close to the same age as I am, takes the lead. I'm fascinated with her full sleeve of tattoos and her beautiful, lustrous black hair as she demonstrates with blazing speed how to order drinks and then how to ring food in, how to modify something in the event of an allergy, how to separate checks—

pretty much any scenario we can imagine—all within the first five minutes. After having seen it work, I do recognize the system—it is indeed the same as before—but, go figure, despite Dan's comment, it has of course been updated nearly beyond recognition.

"All right," Rikki says, "let's see if you're in here. What's your number?"

I again flare with joy in anticipation of her saying Cool! or something, but she doesn't react at all when I tell her.

"That's just your temporary number," she says. "Everybody gets that until their paperwork goes through. How about you, Hans?"

He tells her as I emotionally fall away, my pride feeling like it just got booted from the plane without a chute.

The rest of the day goes by swiftly. We do some role-playing where one of the pair of servers acts like the customer and the other greets her and takes her order, and it feels a little silly but it's necessary. The dining room looks amusingly like an Ionesco play with only six or seven single diners all at separate four-top tables, toasting imaginary drinks to one another as their pretend servers describe and then deliver pretend dishes.

It's approaching four and we were supposed to be done by then today, but instead of cutting us loose, Dan has us break down the entire dining room, as apparently there's some kind of event in here tonight.

Yet again, I find myself in that embarrassing, sheepish place where I really need to run, literally at this point—my car is so far away and this building is so vast—to make it home so Jenae can make it to her work event.

Fortunately, they're all pros at this kind of thing, and to prepare the room for a private function tonight, we fieldstrip the dining room with military precision within five minutes, and I thank Cate and head out the door.

"See ya, cutie," she yells after me, and for the first time since our training began I leave with a little whoosh of joy. I get to the break room downstairs to clock out, proud of myself for remembering this time, but it's packed wall to wall with a beer hall staff meeting.

There's no way I'm going to interrupt that—there are like fifty black-shirted beer hall servers and wait assists crowded into the Mazda-sized break room—and the tablet we clock in and out of is at the far end of that room, and I figure, yet again, I'll have another tedious thing for Dan to do because of me.

Strike 2, squirrelfriend.

Thank God, Emory was better today and hadn't puked while I was gone, but neither of us really wants to send him to school tomorrow.

Jenae levels a look at me like I'm being a selfish asshole by presuming that she can stay and work from home while I do another day of training at Surly. I don't blame her, but I get the sense that she would be totally fine with making these sacrifices if it were for teaching. Even though this is required training, her look suggests, it's just for a restaurant.

"I just," I say, but I don't know what to say. It's mandatory—what choice do I have? "I'm sorry," I say, wanting to make excuses but having none, especially since tomorrow is beer training. But if I'm going to do this job well, I believe I need to be there for as much of it as possible. I feel the need to remind her of all of this, but we've been married long enough and she's expert enough at reading my body language to know what I'm thinking and will mercifully save me from messing up what I mean by saying it out loud.

"I guess," she says, letting the conversation off the hook, "it'd be different if he had been running a fever during any of this. If it acts like it was just a stomach bug or something he ate, I suppose we treat it as such. Hopefully he'll be better tomorrow and can entertain himself so that I can at least get some work done."

"Thanks, hon," I say, drawing the first full breath of the past couple of days knowing she's got us all covered so that I can keep getting up to speed with Surly.

After she leaves, I plug Emory into a *Phineas and Ferb* marathon and go to my desk, looking for some blank cards so I can at least begin to study the menu items, but all I seem to have are pink Post-It notes, and I figure what the hell and fold a bunch of them in half and begin scrawling the ingredients on one side and the names on the other. There are only twenty dishes, but their names are still inscrutable (tea egg, guinea hen, hamachi collar) and mostly tell me nothing of the nearly endless lists of ingredients, and after I get Emory to bed, I go through them several times, and then when Jenae gets home I give up and watch an episode of *No Reservations* with her.

As Bourdain boozes his way through Singapore, it strikes me as notable, if not profound, that I've been immersed back in the food and beverage industry for about a full week now, and the first thing I turn to for entertainment is a show about eating and drinking.

It's not a heady documentary about the Fugitive poets or a biopic about Emily Dickinson or even the Leonardo DiCaprio adaptation of *The Great Gatsby,* the disc of which we've had on loan from Netflix for over three months. I mean, I have a PhD in creative writing and literature, but the last thing I grab for entertainment is a film about things literary or a book itself.

Maybe I'm back in the right place after all.

The next morning Emory is up and ready to roll and go back to school, and both Jenae and I are so, so grateful for so many reasons. When I arrive at Surly, nearly everybody is already in the private dining room again as we were on the first day, and everybody's stack of flashcards seems to have grown overnight compared to my puny little batch of pink Post-Its. Esme's, however, is still the most impressive, and I hope that by sitting next to her I can glean some knowledge by osmosis or something. She seems Spockish and unamused when I ask her if she made her flash cards—*of course I made them,* I can see her think—but what I mean is the paper itself. It's artisan looking, with deckled edges and at least three or four unique sizes and colors, textured too with a nice grain. But when she says simply "yes" to my question, I figure I best let it drop.

When we're just about to start, Cate flies in, almost late, and sits next to me, and my spaniel heart practically leaps in her lap.

"Hi, cutie," she says under her breath. I can't help but blush, but then she says the same thing to Rae, and I realize with a bit of disappointment that this is simply how Cate sees the world. In her eyes, we're all cuties and squirrelfriends. Even Rae, who is, objectively, not all that squirrelfriendly in her angel of death all-black sackcloth getup.

I imagine there's a way to regard Cate and her constant cutie-ness with a bit of skepticism. I, however, do not. She just seems like an absolute force for good, but neither does she appear to be ready to take anybody's bullshit. She's a few inches taller than me, a bit more muscular too, and she shaves her head like a cage fighter. But she is also sharper and more open to share her thoughts or feelings than I ever have been. I've also come to learn she's about to graduate from the University of Minnesota with a degree in comparative literature and is basically as smart and articulate as any badass literary or cultural critic you could pick. I think of her as a Simone de Beauvoir or Hélène Cixous who just also happens to like drinking beer and using the word "cutie" a lot.

Gary is at the head of the table, futzing with his laptop as though he's making a few last-minute tweaks to his presentation, but I get the sense that he's really just keeping himself occupied so he won't have to make small talk until the day begins. I know that trick too.

"Hey, girl," Cate says to Rikki.

"How you doing, lady?" she asks with a knowing smile.

"Oh," Cate says, "mama got turnt last night! Perfect setup to beer training!"

"On that note," Gary interrupts, "I think we're all here and we've got a lot of beer to taste, so let's get under way. Also, let's be clear," he says, giving Cate and Rikki wizening looks, "this is not a beer *drinking* class. This is beer *tasting*. I encourage you to use the empty glasses in front of you to spit out the beer after you've tasted it so we can all stay sharp."

As though on cue, Dan rolls a cart laden with fifteen glasses of what would at any other time of day be a beautiful sight to any beer drinker but at just after nine o'clock in the morning feels a bit like punishment, hangover or no.

"This is Mild," Gary says, and we all start rifling through our dense packet of beers, but nobody seems to be finding what we're looking for. "I apologize," Gary says, peeved at our diverted attentions, "but there's no cut sheet for this one."

"Gary," Cate teases, "why you got to do us like this? I fucking hate Mild."

This, clearly, is an earned tone to take with Gary. An indefinite amount of time will clearly need to pass before I use one anything like it, but I am immensely relieved that somebody is able to take an arch approach with him. It's not that I don't like him. I respect him very much. I just don't know what's on the inside. I try to fix on the fact that he's got a really nice Co-Motion touring bike that he rode to work the other day and not on the fact that he has all of the conversational polish of Data from Star Trek.

Without any more prelude, he begins, completely ignoring Cate's surliness. "The hops are Warrior from the Pacific Northwest, and the malts are Medium Crystal, Coffee Malt, standard Two Row, the yeast is English, and for an aromatic element there are oats. Now," he says, "what's the most important thing about tasting beer?"

It all seems obvious, except for a full couple of beats, nobody answers.

"The smell," Hans says, finally, with confidence as usual.

Gary smiles as broadly as he is able. "Exactly. The first part of anything we taste is with our nose. What's next?"

"Our eyes," Rae says, eager to be a good student too.

"Yes and no," Gary says.

He seems to want to be unpredictable and maybe even a little contentious. This doesn't seem to sit super well with Rae. She is no pushover, and I get the distinct impression that she works hard to be right as often as she can.

"We do use visual data to evaluate what we think of food or beverages," Gary says, getting comfortable at the pulpit. "But," he says, "we can still get the full flavor of food or drink without seeing it. If you have a cold or a sinus infection, I may as well have set down a glass of Old Milwaukee in front of you because without your sense of smell, it'll taste about the same."

We continue our discussion of Mild for a few minutes, and some of the words I jot down as fast as I can are caramel, yeasty, cereal, bready, biscuity, silky, mildly astringent. They're not Gary's words—he's letting us tell ourselves and one another what the beers are—but neither are they mine. If they were my words it would've simply been "beer-y." I'm just writing things down as quickly as possible so I can keep up with these pros who have obviously been around this beer before. Of the dozens of Surly beers that have been available over the last decade, I've probably only had three of them, and for the most part, I only really like Furious. But, for the sake of art and science and a job, I am willing to learn.

"Next up," Gary says as Dan again rolls in a cart full of lovely beers, "is Schadenfreude. Who can tell me what Schadenfreude means?"

I can't help myself on this one.

"Watching somebody in a Ferrari fail at parallel parking," I say, thinking back to one of my fondest memories while having lunch in Aspen, but Gary's unimpressed. I gave an example, not the definition.

"Taking pleasure in another's misfortune," Rae says, smiling her sharp-toothed smile.

Gary repeats what she said verbatim to emphasize the fact that she said precisely what he wanted her to say.

"No brand sheet here either," he says, halting our paper shuffling again. "Sterling hops. Munich malt, both light and dark. Lager yeast. Who smells what?"

Banana bread. Burnt sugar. Sour melon. Light citrus.

These guys are good is all I can say. "Beer" is, again, all I've got.

"And what about food pairings?" Dan chimes in. "That's the other important part of what we're doing today, as well as tomorrow and Saturday."

A couple of the servers lob thoughts about the Schadenfreude going well with the trout roulade or the guinea hen, but I'm mentally stuck on the fact that he brought up training again for tomorrow. I for sure replied to Dan's email about availability that I couldn't come on Friday— I've got a conflict with the other place. That ugly, sinking feeling is creeping up from the soles of my feet like I'm standing in rising water that's about to soak through my shoes before it works its way up my legs and beyond.

But there's no time or space to bring this up now, and when Dan goes to get another cart of beer, Remi and Cate and Esme all leap up to start clearing away empty glasses so we'll have room for the new ones.

"Now," Gary says when Dan rolls a new cartload in, "Warrior and Willamette hops, Two Row, Medium and Dark Crystal, Chocolate malts, oats, English Ale yeast . . . this is where it all began."

All of the pre-indoctrinated downstairs servers say at once: "Bender."

Several folks were clearly unimpressed with Mild. Schadenfreude was well received, but reservedly so. It is an "interesting" beer, just not one anybody is going to order two of. With Bender, the tone in the room palpably brightens.

"I *love* Bender," Cate swoons. She speaks softly, intimately, as though only to me, it feels, but I think pretty much everybody can hear her.

"New Year's Eve," Gary says, "Omar started what would become Bender in his driveway in Brooklyn Center in 2004 and finished the brew the next day—the next year—at three o'clock in the morning. It was from the first exactly the beer Omar wanted to make—a complicated but lovely beer on its own, as well as a beer that would serve as the base for extensions such as Coffee Bender or Cacao Bender."

Gary needs no prompting from his laptop for any of these details, but with the other two beers the details felt academic and perfunctory. With Bender, Gary's tone, and everybody else's too, has become warm and affectionate—reverential too.

"I suppose it's worth noting," Gary says, "without Bender, none of us would be here."

"You know what Bender has always reminded me of?" Cate begins, musing into the caramelly, brown glass she cradles in both her hands. "Those chocolate oranges you get on Christmas—the ones you smash with your palm and they split open into those perfect, gorgeous individual slices. I always loved those and that's what Bender reminds me of."

That, I think, is the most brilliant appraisal of the taste of anything I have ever heard. It's subtle, it's specific, it's nuanced, it describes while it tells an emotional story. I have got a lot to learn.

While everyone is thinking about how brilliant she is, something about what Cate said has triggered a switch in my imagination, and for reasons that are more instinctive than intellectual, I say, just barely audible, "guinea hen."

"What cutie?" Cate says, but, like we all do, the asking of the question was really just a way of processing what she heard. "The guinea hen? Oh! You guys! What Matty said! The guinea hen would be kick-ass with Bender! There's something . . . what is it? . . . oh! The Thumbelina carrots! The pea and mint puree! The way the roulade is cut—all *Game of Thrones* style . . . I know . . . I can't help it—it looks like a severed limb, which always reminds me of *Game of Thrones*. Anyway, Bender and the hen. Fuck yeah."

Across the table, Miko, our mustachioed So-Cal bartender, is nodding. "Totally," he says, "For sure." Dan and Gary make eye contact and both smile—reservedly—but still.

I am happier in this moment than just about any I've ever had in a classroom.

Shortly after Bender, we all go downstairs and have lunch, and I sit across from Cate and next to Rae. Rae gets the salmon salad, and both Cate and I get the hot link (I'm only symbolically vegetarian, and I don't think meat in tube form should count—or bacon). We're all kind of woozy from sampling so many beers on empty stomachs, and when our food comes, everybody pretty much dives face-first into their plates.

Somebody passes around a little cast-iron skillet with the Brussels sprouts, and I spear a bite and can't help but exclaim, "Oh my God—that's better than my wife's!"

"*That*," Rae says, as sharp and judgmental as Johnnie Cochran who just caught a racist cop in a lie, "is the second time you've said that."

I am instantly filled with remorse and embarrassment. Cate doesn't pay any attention to anything except her hot links—she's almost finished with her second one and I'm only partway through my first—but Rae smiles like she knows I know and that means she doesn't need any kind of audience to have heard. She's got me.

At what, I'm not sure. A stupid thing to say? A vaguely misogynistic thing to say? A specifically anti-wife thing to say? Something that could be construed as flirting, but with whom? The Brussels sprouts themselves? Cate, who loves everybody and is loved by everybody in return?

The truth is, I have no idea. I've never said such a thing before trying the food at Surly. Honestly, I would say to Jenae, "I might like these Brussels sprouts better than you, hon. Sorry."

I think she would understand. I think she would say the same thing about me if she'd have tasted them. Maybe even our son. The food here—it's epiphanically good.

But I also understand why Rae seems to be reveling in catching me at something. It was a stupid thing to say once. It's unforgivable to say the same thing twice, especially given the same audience. I use words for a living—I'm a goddamned published writer—I have a Ph-freaking-D

—I'm a tenured associate professor of English. What the hell is the matter with me!? Why can't I use my words here?

I tuck myself back into my plate and try to think of a way to explain to Rae or Cate or anybody who will listen what's the matter with me.

But I don't say anything. I don't know what to say. I'm so ashamed.

"Look at you, squirrelfriend," Cate says. "I totally plowed through my whole plate before you've even finished one link!"

"I know," I lie. "I had a big breakfast."

Back upstairs, we sit in the dining room, waiting for Chef Jorge and Joanna, our pastry chef, to tell us about the desserts. Rae passes around a crib sheet she made for the food—she even made copies for all of us—and everybody's saying thanks and also how we need to come up with a similar one for our beers so we can have both in our server's book when out of nowhere Jorge barks something so rough and loud we think it's a joke. We are wrong.

"I said," he repeats, "how about I talk now so we're not here all fucking day, or do you want to talk and make us all wait?"

We had all been talking like a bunch of chickens—not really one of us louder or more distinct than anybody else—but he decides to level his comment straight at Rae.

Her eyes smolder like a tire fire, but she otherwise does her best to grind her anger back. "No," she says, her voice shaky, barely under her control. "Please," she says, "you go ahead."

The rest of the meeting goes by without any more flare-ups, but I know I feel terrified for even being at the same table as Rae. There's nothing new about any given chef being a hothead—it's pretty much right there in the job description—but it's the first I've seen at Surly, and it has left me a little rattled. Really, I have to admit that I'm relieved it wasn't directed at me.

Schadenfreude, I think in Rae's general direction.

Not a very nice thing to think, and I wonder if there's also a complicated German word for feeling guilty about taking pleasure in somebody else's discomfort.

Back in the boardroom, the rest of the beers mostly wash right by me. Coffee Bender. Cacao Bender. Pentagram. Witch's Tower. Smoke Lager. Mole Smoke.

They just keep coming, and whenever they have a name that ought to tell you what kind or style the beer is, there's also something to subvert or frustrate its categorization. The Smoke Lager, for instance, is not a light, crisp, sessionable lager like the name suggests but rather a Baltic porter aged on oak with an ABV closing in on most wines.

The only thing I take away from the rest of the afternoon's training is from talking with Levi, another downstairs server who is at least as old as I am. He has a lovely, lush, long beard and hair that rivals any Old Testament prophet. He and I have struck up a kind of humorous competition to pair food with increasingly unlikely and spicier beers, the puya chile–infused Mole Smoke (caramelized cauliflower, the salmorejo ragout rabbit, the pork jowl) or the chocolate-, vanilla-, and coffee-flavored Cacao Bender (the crudo, the hamachi collar, the chorizo- and orange-glazed octopus).

It's like we started the morning thinking that beer is like water and its main function when talking about it with food is simply to wash it down. Where we're ending the day is more like conceptual art. What we're after isn't the representation or repetition of an experience but the creation of something indelible, unforgettable—something heretofore none of us have ever thought of as even possible.

Before I know it, all of my senses are overblown and it's time for me to scoot out of here. I know I am missing out on things by tearing myself away right when everybody is winding down and heading downstairs for our shifties—the free beer we get after every shift—but I have got to get my boy at Running Club and then get home, walk the dog, and then get to the other place for training, but as I hit the bottom of the staircase I notice the gift shop and manage to remember Mother's Day and hop in and buy the first thing I see: a purple hoodie with the Pentagram label artwork on the back. It's a sour ale Surly brews and ages on red wine barrels, but that doesn't matter now. It's the best I'm going to do for Mother's Day shopping—and I now get a discount.

It reminds me in the best way of the Easter when I was eight years old and my grandmother, for reasons I'll never understand, gave me, in one of those proper Easter baskets with the chocolate eggs and the

Peeps and the jellybeans and the fake green grass, the album *Destroyer* by Kiss. Jenae and my grandma got along like a house on fire and, bizarre as the connection is, I think she's really going to like it.

As usual, I end up sitting at the light at University, aging, it seems, a year a minute as I stare at five crows atop one streetlight. It feels like an omen that on any other day I would have taken as portending dire events, but today—today, I don't know. It feels like a perfectly Surly thing to see.

Chapter Five

HUNGER GAMES

AT HOME, I PLOP EMORY IN FRONT OF THE TV AS I IRON MY black pants and white dress shirt—my lame but industry-standard uniform at the lake place—and then wait for Jenae to hurry home so I can rush to work.

So far, Surly has been daunting, scary, intense, and also sublime. I've been as challenged there as any job in my life. Though I had to study longer for my PhD exams, I don't think studying for Surly has been any less intense. If anything, it's been more so, since it's not sympathetic academic folks who will be judging me but rather completely dispassionate diners, coworkers, and Dan and Jorge who will decide if I know what I'm doing or if I, yet again, have my head stuck up my ass—a phrase that would almost never be uttered in the politically corrected halls of academe but can be heard hourly in every restaurant in the world.

The lake place—or the other place, as I more frequently call it—is exactly like 99 percent of restaurants in the world that are doomed to a short term on this earth. I was hired by an enthusiastic, kind, and globe-trotting manager named Doug (when I applied for the job, he spoke Spanish to one applicant, Farsi to another), but he was under the thumb of Dreyfus, the pretentious, self-fashioned Gatsby-esque owner, and his two bail bondsmen/student loan–repo men sons. Dreyfus was really a hairstylist who fancied himself a restauranteur as though managing one head of hair at a time is roughly the same as running a five hundred–seat lakeside concert space, bar, and restaurant.

Turns out they are, in fact, quite dissimilar.

But the job was only steps from my house and they met my own terms (no brunch, no frozen drinks) and they made me a bartender because I said I was. And so I was.

When Jenae arrives, I literally just kiss her on my way out the door like romantically entangled tag-team wrestlers and make the ridiculous one-minute drive to the lake place. It'd only take five minutes to walk there, but the skies have just torn open and a torrential rain has begun to pound down on Saint Paul. As I pull into the parking lot the heavy rain reminds me of what my friend Bruce used to say of heavy Gulf of Mexico storms they'd endure in Houston: "Like a cow pissing on a flat rock." Even with my little umbrella, I'd be soaked to the bone in the hundred-yard walk from the lot to the restaurant entrance, never mind the walk from my house.

There's nothing to be done but wait it out, so I contemplate Mike Doughty's lyrics coming through over the radio as I stare at the water pummeling down on the lake like nickels on a cymbal, "Sheep go to heaven, goats go to hell."

Tonight is what's known in the business as a "soft open." It's not publicly announced and there's no OPEN sign lit up. Basically the owners and managers have invited their friends and other industry folks to come eat and drink for free to help us figure out what's working and what isn't before our real open.

Which is tomorrow.

None of us have tasted the food. The computer system is still in flux. We don't really have a flow of service plan. And there are three owners and two managers and nobody at all appears to be in charge.

Once inside, the restaurant is full of damp employees, and when I slip behind the bar, it's as congested as a cattle chute with me, Glover, Paul, Felix, and Jeni crammed in there. I wasn't scheduled to open today, so now there's already a bit of a crowd, and when I join the other four bartenders it feels more than a little ridiculous. Even if we were completely slammed, there would only be so much the five of us could do without getting in each other's way. Given a bar of its size (twelve seats) there's really only room for two bartenders and maybe a barback every now and then to help out with the dish tubs and glasses—which we apparently have to wash entirely by hand. Even with the fifty seats in the dining room that we also serve, five makes us all feel like we're working in a clown car that happens to serve drinks.

Nonetheless, it's good that I'm here, because I finally get a chance to see how the beer taps work, how we'll manage the glassware, and we all get a chance to make a handful of the signature cocktails—most of which include ingredients that none of us have ever heard of like aquavit or Clarito bitters or violette.

Still, it's not all that busy and nobody quite knows how to appear busy. We all know from working in the industry before that the worst thing you can do as a waiter or bartender is look like you don't have anything to do, but we don't.

The bar just got stocked. All the bottles and taps are new and clean. Everything is new and clean.

Glover, in addition to being a bartender, is also the assistant manager. Jeni says his name is "like Lover with a G," and they know each other from a barista gig at Caribou Coffee. Glover is talking about his girlfriend a lot, trying, it seems, to keep the obviously interested Jeni at bay, but she is also somehow young and cute enough to be above the fray of worrying whether a potential somebody is involved in a relationship. They're both adorable.

Paul's a father and relatively old guy like me, but he's got a newborn at home and has the unburdened-for-now air of someone who just came from the far more demanding world of parenthood and is pleasantly at ease to be doing mostly nothing. He seems to have worked enough bar/restaurant shifts to know that there are times to pretend you're busy and times to acknowledge there's simply nothing to be done. He's allergic to small talk—not so handy for a bartender—but he regards me with a kind of gracious equality, and the silence between us is less stony and more of a quiet indication that we're not so childish as to need to chitchat to understand each other.

Being privately less sure of myself, I mess inside the coolers, marrying the half-empty six-packs of beer for as long as I can before we finally start to fill up and have actual things to do.

Since I didn't open, I'm not responsible for the cash in either of our registers, nor do I have to worry about entering drinks or food in the POS system, but just to get the hang of it I try ringing in a few orders.

It becomes immediately apparent that this is going to suck. Even just getting to the tablet-based POS things is awkward. Dreyfus didn't want any computers or registers in the sight lines of guests, so they're hidden from view, tucked up under the bar and almost beyond the deep reach-in coolers. Jeni can't even stand on both feet and reach one of them.

Felix is a quiet, low-key rockstar—one-half of a local semi-famous blistering punk rock duo—and he's just calm and gracious and unassuming. A chatty couple at the bar knows him from when he used to bartend at a nearby sushi place, and they order a shrimp po'boy and some hush puppies and I try to ring it in for him for practice.

The puppies I find all right under APPS, but when I get to the PO-BOYS part of the menu, each item has two aspects: a description and an icon. There's no icon for the shrimp po'boy, however, and when I finally find the description line for it, its icon says FULL HAMBERGER. We don't have hamburgers, never mind half, three-quarter, or full hambergers, so I'm reasonably certain that the description rather than the icon is what we should follow, but when I go back to the kitchen to double-check, some huge guy in a dark apron is standing there, lording over our helpless, bug-eyed teenaged food runners like Cerberus ferociously micromanaging the gates of hell.

I am a full-grown adult—a teacher, a father, even a hockey coach—and here, where all of that means absolutely nothing, I am still a bartender. I was hired in a position of authority and credibility. And yet, before this irate and irrational strange man I have never seen before, I turn tail and run back to Glover.

"Who the fuck—" I start, but I remember I don't know Glover that well and if I can cuss to him as my fellow bartender when he's also my manager. "I mean," I say, "who's the new guy in the fancy apron back there?"

Glover just looks at me as though I asked him to write an analysis of *Paradise Lost*. (He could do it too—Glover's a poet on top of being beautiful and kind, and he also worked at a local bookshop for a while—but now isn't really the place).

"Oh," he says, understanding finally, "he's some consultant—he keeps telling Dreyfus and Doug and me that we shouldn't open tomorrow—that we shouldn't open for probably another three or four weeks what with all we've got to work out. I guess he won or was nominated for a James Beard award or something. The chef from Corner Table I think."

"Super," I say, not recognizing that this is Thomas Boemer, probably the most notable chef in the entire Upper Midwest, the very same one who runs Revival and has become even more famous for bringing Tennessee hot chicken to Minnesota. "Anyway," I say, "about the po'boys."

Glover says he knows and hands me a legal pad that already has a full-page list of POS corrections to make.

"Okey-dokey," I say, and write down the glitch I found. I do not bother with the misspelling of "hamberger," but my English-major heart breaks a little.

Already, even though we're not actually open yet, I find more than a little distance opening up between me and this place. If I only had this job and not also the one at Surly, I would probably feel differently, but I can't not have had the experience of both of them now, and even though neither is officially open yet, the lake place is rapidly falling down in comparison to Surly.

I've tasted and drank and talked about and studied (or am starting to) every single thing we'll serve at Surly and still have more training to come before we open next week. The lake place is opening tomorrow and I have tasted precisely nothing, never mind having discussed it all in any significant detail with the rest of the front- or back-of-house staff. Surly feels like our restaurant is poised to become something phenomenal, whereas the lake place feels like one of those doomed team challenges on *Top Chef* during Restaurant Wars where the night invariably ends in screaming, crying, and career-ending hara-kiri when the chef getting the boot tries to drag at least one or two other contestants with him on his way under the bus.

Perhaps it's not fair to compare them. Perhaps I should just regard them as they are: two restaurants that happen to be opening in the same market at the same time. Perhaps it's as unfair as comparing two human beings or two animals that happen to be in the same genus and species.

But I can't help myself, and right now the lake place seems like a three-legged mule and Surly a stallion and they're both about to enter the Kentucky Derby. Right or wrong, both restaurants have the expectations of great things from both the community, the owners, and the investors, as well as the employees, but the investment in the lake place already feels delusional. On the other hand, the investment in Surly feels like so sure a bet as to risk insider trading.

Even just looking at the staff is enough to illustrate the profound differences in the restaurants.

Take me, for instance. Doug hired me as a bartender without really talking to me at all about my bartending experience (which was basically none, unless you count that one afternoon at that place in Boston when Rock, our regular Iggy Pop of a bartender, was too wrecked to even pour tap beer). Dan at Surly, on the other hand, wanted to know which was my favorite Surly beer and why, and then he wanted me to talk about styles of service and hospitality philosophy and food and beer pairings and what my reasoning for doing so was. Training at the lake place was largely optional and included lots of stimulating icebreakers such as "Tell everybody your name and what time you got up today." Training at Surly will stretch to two weeks and feels poised to continue long after we're open.

Looking at the rest of the staff at Surly, I already feel like I'm the guy with the least amount of game, despite the fact that I think I'm the only one with advanced degrees (which is an obnoxious thing to say and really pretty meaningless when it comes down to simply how well any of us can do our jobs). The rest of the staff at the lake place, however, is a ragtag enough bunch to make the Bad News Bears seem like major league pennant contenders.

Don't get me wrong. I admire and have a good deal of affection for individuals at both places already. But so far everyone at Surly—*everyone*—seems as authoritative, knowledgeable, motivated, and smart as any food experts I've ever known.

I don't mean smart-for-a-truck-stop busboy. I mean wicked fine-dining smart. I could easily see any one of them working at pretty much all of the best restaurants I've ever eaten at. Some of them, like Cate and Esme, are so sharp and talented I could imagine them hosting their own shows on the Food Network or reviewing restaurants for pretty much any big paper or ghostwriting award-winning chefs' cookbooks.

The folks at the lake place either have never been in the food service industry (Jeni, Glover, and, as far as I can tell, all of our teenaged servers and food runners) or have been out of the game long enough to make it curious and a little dubious for them to be in it at all (such as me and Paul). They might be kind or sweet or cute or young or have other talents that speak to the strength of their character, but I'm pretty sure that's not who you hire to open a restaurant. I mean, that's who you hire to take a risk on once you've got a restaurant up and running. To open a place, it seems to me you want a bunch of folks who know precisely what the hell they're doing. That way, even if the kitchen or the POS system or the tap lines or the smoke alarms all go to hell, they still know how to take care of people. With so many folks new to food service—or, perhaps worse, the few of us like me who have been away from it for so long—our crew at the lake place . . . I just don't know. It feels not only destined to fail but like it was *designed* to fail. Almost like it's some huge conspiracy to defraud investors or lose a bunch of money so the city has to buy out another million-dollar breach of contract (which is what I heard happened to the last folks who tried to run a restaurant in the space the lake place now finds itself in).

All I really know is that there is no way we should be open for business yet, and instead, in under twelve hours, we'll open our doors not only to the general public but also to a party of three hundred or so high school soccer players and their parents. They've already paid for their food and drinks so apparently there's no going back now.

Mike Doughty's lyrics keep coming back. Sheep go to heaven, goats go to hell, goats go to hell, goats go to hell . . .

The next day, even though we're slated for our big opening night at the lake place, while taking Frankie for a run, all I can think about is Surly. For starters, for the first time since grad school, I've finally made proper note cards—the names of the twenty dishes on one side and my boiled-down descriptions of them on the other. This was no small undertaking. For every dish, Chef gave us long spec sheets replete with the name of the dish, its global provenance or inspiration, its typical ingredients, and then Surly's whimsical take on them, and then the litany of various sauces, gelées, creams, crèmes, dashes, swooshes, spots, dots, garnishes, and micro-garnishes.

After several hours of study last night and this morning, about the

79

only dishes I have down pat are the beef heart Reuben ("This dish starts with a Wagyu beef heart cured in a pastrami brine for a week, and then it's smoked all day and sliced whisper thin and served atop a house-made charcoal bun with a caraway kraut and a Mahon cheese fondue") and our cauliflower ("Probably my favorite dish on the entire menu—and I'm only a part-time vegetarian!—our cauliflower is caramelized, tossed in a spicy Thai vinaigrette with a house-made fish sauce, and then served over a brown butter and cauliflower puree, as well as a carrot and peanut puree"). The rest of the dishes and their numerous ingredients and their inspirations and whatnot float around my head, untethered from one another like words overheard in a foreign train station and that I'll just be lucky enough to spell or pronounce, never mind attach to a dish. The duck tongues (still can't believe that's a thing), the lamb sweetbreads (can't ever remember which gland that is), the veal heart (are we really serving two heart dishes?), the crudo, the boquerones, the something gribiche, the salmorejo ragout—they just go on and on and on. I love it, but I am so far way outclassed by this chef and his menu.

As Frankie and I turn around and begin our approach of our neighborhood lake and the new restaurant, a name pops in my head from out of the blue, but it has nothing to do with the lake place. It's for the restaurant at Surly.

The only thing that has lacked luster at Surly so far is the name for our restaurant. "Oast House" is what they've got for now, but nobody is even willing to put it on paper. We refer to ourselves the way novelists with bad working titles refer to their drafts. Instead of saying, "So in *How Terrible the Torrent of Tumult* . . ." a writer would say, "So in my book about horse thieves . . ." Likewise, we're just "the restaurant upstairs." Doesn't exactly have a ring to it, but anything is better than saying the word "oast."

First of all, nobody—none of us—knew what the word meant. Apparently, we're told, they're these conical hut-like structures in Europe where they used to dry hops for beer. It kind of goes with the fact that the restaurant is upstairs, and obviously the connection to beer couldn't be clearer, and, though it's a stretch, you can see the Witch's Hat from our patio, a conical water tower in Prospect Park—but still. Oast House? It feels like the kind of thing a toddler is demanding through a mouthful of tapioca pudding when he wants to go outside.

As we run past the lake place I see some folks inside I ought to recognize but don't, and I just can't help but think that, unless we want to explain what the hell an oast house is to every single person we ever talk to at the restaurant, we've got to go with something else.

Sickle. Scythe. Sickle or Scythe. Those are the words that strike me on my run, and as soon as I get home, I towel off as much sweat as I can and hop to my computer and start in on an email to Dan.

> Hey Dan,
>
> In the event that we still don't have a name—I don't know if I mentioned this, but I'm a writer and get positively obsessed with this kind of stuff—and if anybody's still interested, I was thinking that Scythe (or Sickle) would be a cool name for either a Surly beer or, hell, the restaurant.
>
> My thinking is that 1. it's a tool used for harvesting grains—maybe hops too, therefore it's a tool for both food and beer reaping and 2. like our restaurant at Surly, it's an improvement—an elevation, if you will—of an already existing model (in the case of the scythe, it replaced the sickle, and it allowed the user to harvest standing up—cool, right?), and then 3. it's a little bit heavy metal what with the associations with the Grim Reaper, and while I imagine that won't be for everybody, my money says it goes pretty damned well with the Surly attitude and beers (Darkness, Pentagram, Blakkr . . .). And if anybody thinks it's too evil, we can always steer them back to the fact that it's a lovely, almost perfect farm implement.
>
> Scythe at Surly. The Surly Scythe. Or just Scythe?
>
> Far as I can tell, there's not one restaurant by that name in the country, just like us.
>
> There is, however, a beer called the Scythe and Sickle by the Ommegang Brewery in NY state—an Oktoberfest only regionally available I think.
>
> Anyway, hope it's not too annoying that I can't keep my mouth shut.
>
> Cheers,
> mb

My finger hovers stupidly above the mouse as I debate whether I should further embarrass myself by sending this idiotic email and risk proof-positive evidence that I am a moron and have no place at this job, but the temptation is just too great for me to overcome and I hit send.

Even after I send the damn thing, I can't help thinking about Surly.

Unlike feedback on the name, Dan actually did ask everybody to come up with a list of ten songs to send him so he can work on a playlist for the restaurant, and this is the kind of invitation I take straight to heart. So here goes, I figure.

Beck's "Devil's Haircut" leaps to mind, and after that I am off to the races.

> Flaming Lips, "Do You Realize?"
> Arcade Fire, "Neighborhood #1"
> Spoon, "I Turn My Camera On"
> The Shins, "New Slang"
> Wilco, "Jesus, Etc."
> Iron & Wine, "Naked as We Came"
> Andrew Bird, "Fake Palindromes"
> The Decembrists, "Oh Valencia"
> Death Cab for Cutie, "I Will Follow You into the Dark"

This I could do all day, continuing to revise and rethink and contextualize, but I don't, and I'm proud of myself for my impressive self-restraint. I hope in some small way it'll serve as a reprieve for my other geeky and unsolicited email, but I doubt it and send it anyway.

Shit. I finally look up from my computer to see that I've got to get my ass in gear. It's already two and I need to get something to eat before I have to be at the lake place for opening night at four. I quickly shower, put way more gel in my hair than usual, throw on my white oxford, and rush over to Jimmy John's for a sub before my first real shift begins.

I can't help but think, as I cram down my lousy "Beach Club" sub, what the hell am I doing? Not just eating at Jimmy John's (which, I think we can all agree, may as well just make extruded pink nutritional slime and skip the whole pretense of serving actual sandwiches).

I mean with my life. What the hell am I doing, about to work in not only one but two restaurants?

But I don't have time to get philosophical. Not only do we not have any savings but we don't even have a savings account, and our checking account is only a rumor away from being overdrawn. Things got slightly better after my son went from daycare (nearly a thousand dollars a month) to grade school with aftercare (now only almost three hundred dollars a month), but the clutch just went out in my trusty old Subaru (adiós six hundred dollars), I owe sweet Sallie Mae another seven hundred dollars this month, and we're twenty-nine days late on this month's mortgage, which, thanks to our less-than-stellar timing when we bought our little fourteen-hundred-square-foot house, amounts to almost two grand.

Even my shoddy English-major math can calculate that not only do I need this/these jobs, but I need them right freaking now.

I choke down my sub and head in to the lake place for the big opening day. Inside, the atmosphere is calm but jittery. It's all hands on deck again today, which makes the restaurant seem a bit crowded again despite the fact that the doors aren't yet open. We are, however, already busy because of the three-hundred-person ticketed event on the pavilion out back for the soccer boosters. As far as I can tell, Chef Cerberus was right: opening today is a terrible fucking idea. The paint in some places is literally still wet. Still nobody knows how to use the POS system with any confidence. The dishwasher hasn't shown up yet. The printers aren't working because the Wi-Fi that connects them to the POS system is down. The only staff who appear to be old enough to drive here on their own are me and Felix, and I'm pretty sure that we're also the only ones today with any restaurant experience at all (and mine was over a decade ago and Felix has been out of the game for a while too, having been doing, in addition to aggro punk rock, software development in Japan for the past two years). On top of it all, in this restaurant with two bartenders, two counter servers, five food runners, and an expo person, I'm the only one who brought any pens. Within the first hour I'm down from six to one, and then they're all gone by the time we're in full dinner rush mode.

It is, as the industry parlance goes, a total shit show.

The owners and managers never considered the fact that the kitchen is roughly the size of a Honda Accord (the two-door model, not the roomier sedan) and every dish that comes from it has to go through a window the size of the slots in prison doors via which they deliver meals to those in solitary confinement. The dining room seats only fifty, which would have been fine, but we're also taking orders for anybody who wants to sit on the pavilion out back. Which seats about five hundred. For better or, likely, worse, at the bar/counter, we're soon able to take order after order after order with little to no problem on our end, but what we don't know is that the ticket times (how long an order goes from being fired to being delivered) have scooched up from fifteen, to thirty, to sixty-plus minutes.

But since there's virtually no waiting to put your order in, guests expect the food will arrive shortly. And then the food doesn't show up, and it doesn't show up, and, in a special few cases, it never does. Because of the fun and modern (and cheap and secondhand) POS system, sometimes the orders we send to the kitchen simply evaporate. The super fun part for everyone involved is that nobody knows this is the case until the guests come back red-faced and frothy after an hour and a half of waiting with food nowhere to be found.

Total. Shit. Show.

To everyone working, whether the beleaguered and doomed cooks, the frazzled first-time restaurant folks running food, or even us slightly more seasoned bar staff, it is a clear and complete failure.

To the owners, on the other hand, there are people in their restaurant spending money—Dreyfus and his sons are literally puttering around the place, just staring at the app on their smartphones that tracks how much money they're making. To them, it's a smashing success, and they're quack-laughing like Scrooge McDuck and his nephews as they prepare to bathe each other in gold coins.

In the middle of the very next shift the very next day, we're busy as can be again, up to our necks in irate guests, and it feels like it will never get any better. It's Felix and me tending bar, with Doug and Glover floating around trying to mollify the angry hordes, problem-solve the persistent Wi-Fi failures, and train their teen food runners how to carry more than one plate at a time. Jeni is on the other side of the bar. Having just finished her lunch shift, she's perched on a barstool, mocking me and Felix as we keep running into each other and spilling our drinks all over like a couple of vaudeville mimes. I glance up during a momentary reprieve after having set down a pair of cosmos in front of a couple at the bar just to see Jeni's eyes follow Glover lovingly as he run-walks toward the kitchen when she stops watching him and pauses, mid-sip of her beer.

"Do you hear that?" she says, settling her drink on a coaster.

"The sound of despair?" I say, rolling up my sleeves, readying to tackle the sink full of dirty glassware. "Feel free to jump on back in."

"No," she says, "I mean, sure, it's that. But it's something else."

It's weird that I don't know what she means. Despite the fact that the restaurant is packed, it's relatively quiet, especially since Dreyfus and his sons were too cheap to pay for a sound system, so there's no music, just the vulgarities regularly floating out from the kitchen and the din of grumpy people in the dining room not getting what they want. But then I think I hear what she's talking about.

"Is it, like," I start, "water?"

Jeni's young—probably only in her early twenties—but I get the impression that she's seen some shit in her years of being a barista. She had changed out of her work clothes into something apparently ready for the club—a miniskirt, a sequined tank, and suede boots—but now she starts gathering her hair into a high ponytail and hops down from her stool to check things out.

"Oh," she says, only having got a couple of feet. "It's water all right."

I follow her gaze to where hers lit on the bottom of the stairs, and there, sure enough, is a little trickle of water. For the moment, I don't think much of it. The owners had renovated nearly everything in this century-old building, and I imagine they just didn't think through letting guests take drinks to the second floor. Somebody's just knocked over their gin and tonic is all. Maybe a pitcher of ice water. No big deal compared to all the other nonsense going on, I think, when Jeni begins to take off her suede boots.

"What are you doing?" I say, thinking that maybe they're just uncomfortable, when, not answering me, she reaches over behind the bar and grabs one of the tall plastic trash cans and heads for the stairs.

Just then the trickle begins to turn into something more like a little stream, and in another moment water starts cascading over the second-story balcony directly down onto everybody at the bar. I start hopelessly toweling off the bar, telling everybody I'll refresh their drinks in just a minute, when suddenly there's a geyser-like release of water from the balcony and all at once every single patron at the bar is soaked. I drop the now drenched towel on the floor only to hear it splash into water that has begun pooling at our feet.

"Battle stations!" Dreyfus declares from his corner table where he's entertaining a couple of his friends. They are apparently dry enough—and insured enough—to find this all amusing. Felix and I just shrug at each other and squish our way to the stairs to see what's happened as all of our guests flood out of the doors. What was a minute ago a trickle is now a proper waterfall across the entire width of the staircase.

Doug materializes with a mop and bucket and shoves them toward me. "I'm going in," he says, turning his apron backward so that it's something of a shoddy cape now.

"What?" I start, but stop. There's no question I can come up with that would cover all I want to know.

If we'd already had the soft open, then this, without a doubt, is the hard, hard open.

Apparently a four-inch water main that pumps hundreds of gallons of water a minute has burst. Felix, Jeni, and I all spend hours—literally—mopping up shitty water while the now barefoot Dreyfus, in his purple velour blazer, mixes up cocktails for himself and his friends. Holding their martini glasses by the stem so as to keep their drinks properly chilled, they all appear to regard the whole scene with the kind of be-musement and indifference that preceded the sinking of the *Titanic*.

"Just keep it away from the walls," he says, taking a sip directly from his shaker. "It'll be fine, fine."

But this restaurant is not the *Titanic*, nor is it the *Queen Mary*. Not even *Boaty McBoatface*. This is just another shitty restaurant undeserv-ing of even an apt nautical metaphor.

Some successes are boring, while some failures are interesting. The lake place, the summer will prove, is an entirely uninteresting failure.

Chapter Six

THE PREGAME
GAME SHOW

I RIDE MY BIKE TO SURLY AND GET THERE A SOLID TWENTY
minutes early so I can stop sweating before I change into my work
clothes for our own soft open. It's only a three-or-so-mile ride, but very
early in my time at Surly it became evident that there were two kinds
of employees: those who ride bikes to work and the even stranger rest
of them. The general staff vibe at Surly skews young, to be sure, but age
isn't the most significant determining factor when it comes to getting
a sense for who works there. I would call it bike messenger/tattoo par-
lor/heavy metal roadie chic. Which is to say about 90 percent of the
staff at least talk about bikes and 100 percent of them have messenger
bags, even if they don't ride, and pretty much everybody who works
there has some kind of ink on their body—most highly visible too, even
a few with neck and face tattoos. Those who don't have tattoos or bikes
aren't the outcasts though. They seem somehow more punk rock for
their lack of striving to conform in this stridently nonconformist tribe.

Given that I'm literally twice as old as some of the beer hall folks,
I hope riding to work will help curb my downward-trending cache. It
also might not hurt that I ride a fixed-gear bike that I built myself—
even the actual wheel I built from parts (with a lot of help from my local
bike shop)—not that I can say that to anybody as an icebreaker but it's
nice to have in my pocket for when chatting with Miko, an actual former bike messenger.

After I lock up, I stroll casually by the smokers at the bottom of the dock—head nods, an almost inaudible "What's up?"—and make my way through the big, heavy, red steel door at the back of the restaurant that says Employees Only.

I freaking love this door.

It swings open with more ease than you'd imagine, despite its size, and on the other side of it is a world you've never seen unless you've worked in a restaurant. This is not Martha Stewart's or Rachael Ray's sunny kitchen with their fake windows looking out onto their fake gardens.

Open any door to most any other workplace or business and you'll likely as not find a saccharine person paid just to greet you and help you find your way or, at the very least, a directory and a floor plan or some kind of map to help you navigate the place. As with most restaurants, the employee door to Surly is the unapologetic portal straight into the belly of the beast. There is no antechamber or receptionist to our variety of hell, never mind a YOU ARE HERE map.

Immediately to the right are huge power panels with custodial carts and ladders around it. WARNING ARC HAZARD is emblazoned across the power box, and I give it a very wide berth. Though I don't know what an arc hazard is, the helpful little cartoon of a man being struck by a bolt of lightning is shorthand enough.

To the left are racks of wood for smoking meat and pallet after pallet of soda, bottled water, and various other kinds of dry storage stuff like paper towels, napkins, and plastic sleeves with hundreds of to-go containers. Farther down is the cooks' uniform and linen closet, employee bathrooms, and then twenty yards of fire engine–red employee lockers, and beyond them a huge throng of beer hall employees. It's a bristling clutch of black-T-shirt-and-backward-trucker-hat-wearing folks who are on their way either to or from a frenzied shift of serving close to twenty-five hundred people on any given day. Brian, the beer hall manager, is in mid-rage as I try to weasel my way into the men's room without drawing any attention to myself.

"If I see another server or wait assist or anybody who's on the clock so much as glance at your phone," Brian says, "I'll fucking put it so far up your asshole you'll be able to feel it vibrate against your heart. No fucking phones on the floor. Ever." He slams his fist against the wall of lockers, making everybody jolt just a little. "Use the motherfucking lockers!"

Once in the men's room, I sit and pad my sweaty torso with paper towels so I can change into my uniform for service.

NO FUCKING LOCKS ON LOCKERS OVERNIGHT! reads a piece of copy paper taped to another bank of lockers. Between that and the vigorous tone outside, I get the distinct impression that they're running a different kind of operation down here than we hope to do upstairs. Still, we share a space and, for the most part, I dig it. I've always liked profanity.

I start changing into my dark jeans and gray denim shirt when Eric (I think—there are so many beer hall employees) comes in and starts to relieve himself at the urinal. Eric is the kind of guy who talks to people in the bathroom—something not universally true in men's rooms—and Eric is also the kind of guy who likes to do so while making eye contact. I typically share neither of these traits, but I like Eric immensely.

"Mick, my man," he says, craning back over his shoulder. (I let the name slide—a four-letter M name is good enough for early encounters.) "What's up, brother? Ready for a big night? Is this it? Opening night?"

"Yeah," I say. "Sort of. It's the soft open."

"Sweet," he says, giving a little attention to the task at hand.

I finish tying my shoes and meet Eric at the sink, and as we wash up, we both notice a white powder sprinkling the dark concrete floor from the sink all the way to the bathroom stall.

"Is that—?" I begin.

"Fucking classic," Eric beams. He shakes his head as though to say, Don't kids do the darndest things? And then he grabs a couple of wads of paper towels and is out the door.

I wonder just what other narcotic evidence there is in the stall, and since I'm alone in the room now, I decide what the hell and tip the door open with my foot, careful not to get my fingerprints or anything anywhere near what I imagine is cocaine. But instead of a little nest of hypodermic needles or a wad of dope balls, against the handicap grab bar on the opposite wall is every cook's better friend than even Bolivian marching dust: Gold Bond Medicated Body Powder.

I head out of the bathroom, and fortunately the beer hall pre-shift is over so I hustle past the managers' offices, none of which are bigger than any given school janitor's closet; past the sanitation room with all of its tubs of scary solvents and racks of industrial cleaning equipment; and then by the beer hall dishland, where four guys are running rack after rack of dishes, silverware, and glassware through a countertop dishwashing machine at the pace of a top-level NASCAR pit team. And then there's the gauntlet of walk-in coolers for meat and dairy and Lord knows what else that all open out into the narrow hall I'm trying to thread though without incident to get to the service elevator. Beer hall servers and cooks and bartenders are going every which way in every space I've just mentioned with about as much situational logic as a new-comer might observe on the New York Stock Exchange floor. They're a mass of black server T-shirts and white chef's shirts moshing together, only every one of them is wielding a pot or a pan, a thirty-pound rack of steaming glasses, or a whole armful of dirty dishes. I might be a Surly employee, but I know there are limits to where I am permitted to tres-pass. I hit the UP button and age a year with every second I have to en-dure being so useless and out of place down here.

The door mercifully opens and in I go, but just before I'm sealed in alone, Rikki hops in.

"Hey," she says without much enthusiasm. She's wearing black jeans and a black T-shirt and has heavy black eyeliner on and she looks like she just lost a fight during her roller derby bout.

"Double shift?" I say.

"Yeah," she says without making eye contact. "Fuck me."

Even though it's a practically brand-new elevator, it's glacially slow. I, for that matter, am no quicker and can think of no witty retort to Rikki.

"Hey, guess what?" I say. "So my number really is 500."

A beat goes by. Then another. Then the elevator stops but the door doesn't open.

"What?" she says, not tracking at all.

"My employee number," I say, fidgeting with the tails of my un-tucked shirt. "Remember," I say, "when we were training on Micros? You said my number was probably just temporary."

Rikki and I are probably within five years of each other. I'd bet she's the younger of us, but who knows? She's taller than I am. Leaner than I am. Has ten tattoos for every one that I have. She's got dark blue-black hair that reminds me of the color of a gun barrel. She's pretty and she knows it and is generally unimpressible.

"Oh," she says. She does not add, for the record, "good for you, little boy" but she may as well have.

The door opens and she shoots ahead of me through the back entrance to the restaurant.

I guess I can't be friends with everybody.

I've figured out not only what a soft open is but simply that they exist just last week, and here I expect to be thrown into the unforgiving pit of one yet again. As I pause just outside the restaurant's back door, I imagine the kitchen to be in complete turmoil; the dish pit to be quintuple stacked with pots and pans from prepping; the floor to be a chaotic miasma of tables in the wrong place, silverware yet to be polished, glassware in no better shape; and the whole lot of us trying desperately to cram some last-minute specs into our heads about the dishes we weren't introduced to until late last week.

But no.

Dishland is mostly empty save for a few hotel pans, sheet trays, and a couple of pots. There aren't even any dishwashers here yet, nor is anybody prepping in the extra space in the back by our own walk-in refrigerator, which now has a picture of Christopher Walken on it saying "More cowbell." I make my way out to the dining room and mumble "corner" as we're supposed to every time we come into or out of the back, but everything is so quiet I don't exactly belt it out, and Pat nearly bowls me over. She reminds me so much of Pat Benatar I really want to call her that but I'm too scared to.

"Jesus fucking Christ," she says. "You gotta say corner like you mean it. Come on! Say it!"

Like her nicknamesake, Pat is short, ferocious, and mostly feral. And also really talented. "I said say it," she repeats, holding what must be a super heavy sheet pan with several dirty pots and pans on it.

"Corner?" I try.

"Louder!" she says, hitting me in the arm with her sheet tray.

"Corner!" I say, as loud as I dare, fully expecting Dan or Jorge to bust in from the restaurant demanding to know who's screaming in the kitchen, but nobody comes.

"That'll do," Pat says and lets me pass.

I know enough to not apologize and squirrel my way out of the back.

Despite my two unpleasant run-ins so far and my rock-bottom expectations of what soft opens are, the line of chefs is busy but silent save for the Janet Jackson they're cranking out as they finish up the last of their mise en place (fancy French for all the stuff they need to have prepped and ready to go to get them through service—diced onion and chopped garlic, picked herbs and squeeze-bottled sauces, and so on). Instead of the doomed feeling at the lake place on its soft open, I'm getting more of a ready-for-the-party mood as Janet belts, syllable by syllable, "Es-ca-pade" with more than a couple of the cooks singing along.

The floor is a little more intense though, and I get there just in time to catch the tail end of Dan tearing into Rikki.

"No discussion, Rikki," Dan says, smiling but blunt. "The uniform is dark blue jeans. Not black. Go home and change or just go home."

Now, I imagine, as Rikki storms out, isn't the time to ask him if he got my playlist and/or if he had any questions about my name suggestion.

I roll out my nearly floor-length apron—the first time I've donned one other than the cheap white one at the lake place in nearly a decade—wrap myself up once, tie it in front, and look around for stuff to do.

Bella, another one of the servers from downstairs who hasn't yet been a part of training, is apparently going to be running the pass doing expo. Basically that means that the chefs are all on one side of the line doing their cooking and plating thing and Bella/the expo is supposed to negotiate from the other side what will surely be the never-ending line of tickets popping out of the expo printer, calling out how many of what dish to make along with when to do it. She's basically got to quarterback the whole line of chefs, as well as negotiate any unique needs we servers have—food allergies, if we'll make something special for a kid, is there gluten in the guinea hen, and so forth.

Without question, I think it's the hardest job in any given restaurant.

Bella's putting down long strips of blue painter's tape—the favorite label of every kitchen I've ever worked in since you can write on it and it's easy to remove—but she seems a little frazzled, more than the rest of us, and I decide to say hi and introduce myself.

She doesn't really reply in words, so I ask if she's all right.

"I got T-boned on the way to work," she says. She doesn't look up from her tape job and her hands are visibly shaking. This multimillion-dollar operation, and two long strips of painter's tape is what holds all the shit together once the tickets start spitting out, only Bella's so shaky that even this does not bode well for the evening.

"Jesus," I say. "Are you okay?"

"I don't know," she says. "I guess I have to be. My parents are going to kill me though."

"Their car?" I ask.

She nods but still doesn't look up.

I don't know what to say. I want to give her a pat on the shoulder or a hug or something, but I'm still academia shy—meaning any physical contact with a student or coworker has me terrified of a career-ending lawsuit, and many of my new colleagues at Surly are the rough age of most of my undergraduates—so I just give the counter next to her an anemic little pat and tell her to let me know if there's anything she needs.

I make my way to one of two side stations where Esme is going over notes.

"Oh my God," I say, but Esme doesn't look up. She's poring over her nifty color-coded index cards. "Bella got in a car accident on her way to work today. Not looking good for tonight."

"What's hyssop again?" she says, still not looking up.

"Um," I stammer, quite unsure. "Some kind of leaf," I say. "Or herb. Well, I guess pretty much all herbs are leaves—or are they? Was it a root vegetable?"

Here she stops.

"No," she says, indeterminately.

"No what?" I say. "Not all herbs are leaves?"

"Shit," she says, letting her notes fall to the counter, looking around for someone more authoritative to consult about hyssop, still unhearing about Bella. "We are so fucked."

I bide my time at the side station and section out pieces of printer paper like I saw Hans do a few days before, until I've got what will hopefully be enough scratch paper to take tonight's orders on.

I check in with Min and Katie at the host stand to see what my section is going to be, and I find that all eight servers are on, as well as a couple of food runners and wait assists, and I note that my section is only three tables. One round four-top and two four-tops on the banquette. If things get hairy and they decide to break up the two banquette tables, I might have four two-tops and the one round four-top, and despite the fact that it's the same number of seats, it could be much worse than simply three four-tops.

Without overcomplicating it, basically it comes down to the fact that, in general, parties of four or more come to entertain each other. Deuces (two-tops) often are already bored with their own meager company and they dine out to be entertained, usually, in large part by their server. I suffer from the need to over-empathize with people, so if I sense that somebody or a couple or whoever needs more attention, I'll bleed time and energy from wherever I can get it to give them that extra care. Doing that for three tables is usually not a big deal. Doing that for five tables can be a lot—especially if they're all seated at roughly the same time.

But that's the future, I tell myself, not the now. All I have to worry about now is, well, pretty much everything.

Finally, it's four thirty and time for our first official soft open pre-shift meeting. I'm nervous it's going to be something like the horror show downstairs with lots of profanity, chest-thumping, threatening of progeny, and so forth, but Dan begins with some news.

"So," he says brightly with that smart-assy, Alfred E. Neuman smile he's always got on, "we're not the Oast House anymore. That will probably be a relief to some of you. I, for one, am glad we won't have to define what it means a hundred goddamn times a night. Instead," he says, pausing for great effect—and I hold my breath, ready to hear him announce my idea as the big winner. "Instead," he says, glancing my way and then away, "we are The Brewer's Table."

He pauses and kind of bobs his head in expectation, but most of us are just like, "cool." I am a little deflated but partially relieved. I don't really want any special attention tonight. I just want there to be no flood or any other catastrophes, unpredictable or otherwise.

"All righty then," he says, "you guys should've already gotten with Min and Katie about your sections. Proof everything in your section in the twenty minutes we've got and then proof it again. No fingerprints on the silverware, no smudges on the water glasses, etc. Check for gum under the tables . . ."

Cate raises her hand. "Um, Dan, cutie? We've actually never been open. If we've got gum under the tables already, we've got problems."

Dan's grin widens and he blushes ever so slightly.

"But," he says brightly, "it's never too early to check for problems, is it? Also remember they had some event in here last week."

"Touché, baby!" Cate says.

"So," Dan says, shifting gears. "We've got a lot in front of us, but we've already put in a ton of work. Let's play a game for a couple of minutes. A pregame game show, if you will. Who can tell me where our octopus comes from?"

I love the octopus and think I should have this one. It's a super tender dish of five or six tentacles that they marinate in citrus and, somehow, chorizo, and then plate in a little octopus Stonehenge. But where it's from—I know it's Spain, but that feels vague and also wrong—it comes from waters in the vicinity, but I'm not super confident about that either. Does that mean the Mediterranean or the Atlantic or what?

"The water," Cate says, still sassy.

"Okay, bright girl," Dan says, "but where in the water?"

"Beneath the surface, of course," she doubles down, but Esme jumps in to make sure nobody gets in trouble.

"From Spain," she says definitively. I don't actually pump my fist, but it takes a good deal of restraint.

"Dustin," Dan calls across the line. "Give me a garnish."

"What?" Dustin says. He's our second-in-command, technically our chef de cuisine, but he's the same age as Cate—they went to high school together—and he's blond and eager faced but also as intimidating as Jorge given how much responsibility is on his shoulders. He's also got a big *Night of the Living Dead* tattoo on his left bicep that is strangely beautiful and terrifying. "You want a what?" he says.

But Dan is poised, unfazed.

"A garnish—anything—we're playing a little game."

"Whatever, Danager," Dustin says, smiling now, and he tweezers over a little green and purple something to Dan.

I expect Dan to react to his apparently new nickname, but he doesn't flinch and instead holds up the little stem Dustin gave him.

It's clearly vegetal—an herb, a flower—but it's so tiny.

To this Cate has nothing funny to say, and when Esme raises her hand Dan wags his head no; he wants somebody else to answer. I curse myself for not taking a shot with the octopus because I now have no clue. I don't doubt this has been on a dish I've seen, but I must not have gotten close enough to actually have it identified.

Technically we all should know all of the ingredients, but there literally are probably more than a thousand parts to our menu when you count everything that goes into everything. Dustin, too, is now waiting for the answer, and though both he and Dan are smiling, their eyes are flat and expectant, like a couple of the villagers from Shirley Jackson's "The Lottery," cold rocks in their hands at the ready. Jorge comes out from the back and works himself down to the line and stands next to Dustin as well, seeming to have heard or known what was going on, and collectively it feels like an ass kicking is about to go down.

But Claire, also one of the beer hall servers, piques up, takes a closer look, and then turns her head just slightly to the left, like a seasoned speech contestant or otherwise well-trained public speaker. For added effect she tilts her head up so that her chin is level with the counter, and she appears as ready to talk as to do a pirouette. "Micro cilantro," she says finally, smiling demurely.

"Winner winner, guinea hen dinner," Dan says, and everybody lets go of the breath we didn't know we were holding.

"All right, everybody," Dan says, glad apparently to end the game here with a victory for us all but conspicuously amped up for what lies ahead. It is, after all, his first day too. "Let's have a great service."

We all double-check our stations, but we've done our jobs—and nobody's ever touched most of our serviceware yet—and before long we all take our places.

Miko and Remi are behind the bar, aloof, both beautiful in their respective ways, each vaguely engaged in polishing glasses as is the custom of bartenders far and wide. Behind them, dozens of already polished stemware stand waiting on the back bar—tulip glasses and stemmed pints and a weird angular goblet thing for the sours. They gleam white and somehow silver in the bright light, dramatic and sharp against the black of the bar, and remind me of the perpetual readiness I felt when I saw an exhibit of those Chinese terra-cotta warriors.

Hans is pretending to check his station nearest the host stand, but he's clearly just trying to hit on Katie despite the fact that she's frantically trying to get the iPad they're using to assign seats to work the way it's supposed to.

Cate and Claire are coolly chatting about something secretive, and they curl into each other to whisper, then splash back with a laugh, then regroup into another pseudo-subtle bit of commentary.

Claire, from what I gather, is hilarious. She's super effusive with the other female staff—big hugs, butt slaps, over-the-top, full-belly laughter—but except for Miko, with whom she also works downstairs, she seems to distrust all of the male staff members. Newly engaged, I've heard, to somebody the other servers call "Playgirl"—something about him being a centerfold–hot bike messenger, I think—Claire has worked for Jorge before and also for Dustin when he ran Fika, the restaurant at the American Swedish Institute in south Minneapolis, and I expect she knows everybody else's antics all too well. But with Cate she keeps going back and forth between her super erect, prima ballerina stance and a hilariously crumpled kind of Groucho Marx when she laughs. Her hair is the color of a lit match, and she too is about to graduate from the University of Minnesota, with a degree in French literature and education. Like everybody here, she's scary smart.

My section is just two rows of tables away from the big picture window overlooking the beer garden, and I stick to it religiously, polishing every stick of already spotless silverware as though my tenure, and not just this job, depend on it. I don't quite know what else to do, but I am completely confident that polishing silverware will almost always pass as being busy.

Sofia is in the section next to me, and she glances my way as she goes through the same routine. She sidles over, nodding and taking me in, not in any kind of predatory way, but just in the way that folks like her do—people on whom nothing is lost.

97

"So," she says, "where'd you come from?"

I tell her I haven't been in the business for a while, and she seems only mildly interested in my reply.

"I have to admit," she says in quiet confidence, "I am nervous as fuck about this. I've never worked with a menu this . . . involved. I hope I haven't taken too many . . . vitamins."

"You're a pro," I say. "You'll be just fine. And I know what you mean. Vitamins can be really hard on my stomach." I'm thinking we're both talking about those daily horse pills grown-ups are supposed to take that just never hit my gut the right way.

"Yeah . . . ," she says, but in a way that lets me know she and I don't take the same pills. "But hey! You know who else be fine?" she says, shifting gears abruptly, needing or wanting both of us to get more hyped about what's before us. "You and me," she continues. "We both fine bitches! Are you Greek too? You've got Greek hair—kinky and curly and out of control like me," she says. "Like my hair, I mean. Anyway," she says, vigorously polishing the fork she's working on to an undeniable shine, "we both foxy, right?"

"Right?" I hazard.

Sofia is sort of the opposite of Claire and Esme. I know she's not flirting with me—she's probably fifteen years my junior and I'm happily married and who knows what or who she's into, other than vitamins. But I know she's not flirting with me. She's just flirty. I can already tell she'll be a huge hit with our guests. She's just got that perfect server's way where every table will feel like they're her favorite.

"So you be Silver Fox," she says, gesturing with her fork in the vague vicinity of my old-guy salt-and-pepper hair, "and I'll be Little Fox."

"Deal," I say.

Just then the mix of hip-hop, show tunes, and '80s glam pop is cut off from the boom box the kitchen had been controlling, and we all look around for the reason why.

Over by the bar, Dan is messing with his iPod, cueing something up into the AUX port. He smiles that smile of his as he nods and reaches for the volume knob on the wall.

Miko shakes his head and cups his impressive moustache and says, loud enough for everybody to hear, "Shit's about to get real."

Cue the greatest mediocre band to ever make it big out of Sweden—the band Europe—with what else?

Dan sings along with the entire '80s synth anthem, "*Dee-dle-dee, dow, dee-dle dun-dun-dun . . . dee-dle-dee, dow, dee-dle dun-dun-dun . . . dee-dle-dee, dow, dee-dle dun-dun-dun-dee-dun, dee-dle-dun, dee-dle-dun, dee-dle dun, deeee . . .*"

It's, literally, "The Final Countdown."

Chapter Seven

THE GERMAN WORD
FOR LIGHT

ORDINARILY, WITH BOTH THE EAST AND WEST WALLS OF THE restaurant being entirely made of glass, you can look in one direction and watch the day fade over the skyline of Minneapolis and in the other direction see out over the balcony at the massive fermenting tanks in the brewery, glinting industrially like giant yurts filled not with hippies and their llamas but beer. But tonight, not so much.

Tonight—though it's just a couple minutes before five o'clock and it's mid-May and the sun won't set for a couple of hours—tonight, on this late afternoon, the east windows are full of people.

Ordinarily, restaurants tend to be dead in the late afternoon and then folks start to trickle in with their walkers and/or diaper bags and/ or Tupperware containers stuffed with dried prunes and baggies of Cheerios—maybe even some plastic placemats for the extra thought-ful so as to make the inevitable clean up a little less tedious—and then around 6:30 things start to get noticeably more full and then, almost without being aware of it, the place is packed and will stay that way even though you might turn each table in your section a time or even two— and then you're back to where you started around 9:00 or so, except that it's young hipster couples with those ironic-sized glasses without kids and then, depending on how things get dragged out or not, every-body gets to go home.

But this—this is nonsense. A whole wall of people waiting to crush us before it's even time for the nightly news. It's not even time for freaking *Wheel of Fortune*, and yet here is this brood marring all of our just-cleaned windows with their damp hands and shaggy breath.

Jorge is at the expo pass with Bella on our side of the restaurant with all of us jittery servers, food runners, and wait assistants. Behind the pass are the rest of the cooks and, compared to us, they're all preternaturally calm. Most of us are still attempting to cram our heads with the difference between the six different kinds of olives we're presenting to each table upon arrival or struggling to recall what the hell egg gribiche or huitlacoche is and which dish they're parts of, and we just know that the little game of "what's this ingredient" Dan had us playing is only over for the moment. Linda has shown up at the end of the counter just to make sure we're all adequately nervous and that all levels of management are present and perched to watch us fail. The only one not here is Omar, or maybe he is and I've yet again confused him for one of the line cooks or dishwashers.

I am prepared for more tests and quizzes and steps of service reminders—maybe some last-minute big shake-ups in staffing or sections or whoknowswhat—but Dan just smiles his Danny smile over by the bar after he decides we've heard enough of "The Final Countdown" and shouts something motivational but also unmemorable, something along the lines of "You guys know what you're doing. Let's have a kick-ass service. Any questions?"

Today there are too many to list for most of us and the line of impending diners is just too great to deal with, so nobody says anything. There's just too much drama and suspense.

As we finish tidying up our sections, our side stations, and ourselves, and as Dan goes to unlock the door, I am riddled with doubt. I've already got what feels like two strikes against me. What if I mess up big tonight? Will Dan immediately kick me out of the place like he did Rikki? What if I can't hack it physically? Bartending at the other place is one thing, but I pretty much stood in one place there and never had to run or explain food. What if I can't cut it here? Sure, I could at plenty of restaurants before this, but so could people who didn't even speak the language. And yeah, I had a PhD and lots of years of experience teaching, but those two things aren't even remotely heading in the same direction as what I'm doing here, and I am several stripes above any restaurant I'd ever dined in, never mind worked at. What the hell am I doing here?

As I work and rework this one last smudge on a fork, I think about the friends and family I've told about this new job. They always smile bemusedly when I talk about how much I have to study, even after I show them my stack of flashcards thicker than those of students trying to get into med school. They think I'm kidding—exaggerating about how difficult it is or why I'm doing it or how excited I am about it. I can't really talk with Jenae about it because I hardly ever see her now that I'm working most nights and she's working days, but when we do, she just nods and smiles. My son is the only one who seems to think it's cool that I get to wear an apron at work. "Like smocks in art class!" he says.

Nobody believes that this might be the hardest thing I've ever done, and also the most thrilling.

But I'm getting way ahead of myself, I see, working ridiculously hard on this one last smudge on the tines of a fork that turns out to just be my own reflection in the metal.

As people cue up behind our hosts' podium I traipse over to the bar side station and make a mental list of words/concepts/items I've basically never known or heard of or tasted before.

There are the rabbit rillettes with the salmorejo ragout. I know what rabbits are—that's the only thing I know from that whole dish.

Then the little neck clams, which I'm fine with, except they come with a vichyssoise and a sunchoke chip, both of which I've got nothing on.

And then the asparagus. I've got that veg down pat, except for the fact that it comes with juniper skyr, a soft-boiled tea egg, and Banyuls vinegar.

And then there's the beet salad that doesn't come with any lettuce and is mostly several kinds of beets. I've always liked beets, even the canned ones my mom served when I was a kid. But this one has ras al hanout something or other, a buttermilk curd, and then shaved foie gras on top. Not quite like mom used to make.

And finally there's the crudo, which after much study I now know means "raw"—in which language I can't recall—and that should be simple enough, right? It's just a raw fish appetizer, but no. This is raw hamachi, which I'm pretty sure is a tuna, but different from big eye and blue fin and skipjack and so forth—I know not how—and it's "compressed" with guajillo peppers, micro cilantro, and garlic and lime and salt, and then served with spiced pepitas (some kind of seed?) and laid over compressed cucumber and compressed avocado (I'm clearly just reading from my notes now—there's no way I'll be able to articulate any of this shit tableside) and then a masa cracker, fresno rings, pickled jicama, and—finally!—a huitlacoche puree. Whatever any of that is.

And to make matters worse, all of that was only the Snacks portion of our menu. We've also got Veg, Fish, and Meat. And desserts, too, but I haven't even pretended to crack their code yet.

The rest of the menu's impenetrability washes over me with a wave of despair. There's the duck tongue with the tamarind something or other. The fried green tomato with the shockingly simple Frank's Red Hot Sauce (but in what and how, who knows?). The opaque panzenella with boquerones, pincholines, and espellette powder (WTF to all of it). The Reuben. The veal taco. The duck heart with tamari and togarashi crumble (wait—another freaking heart?!). The guinea hen with Fernet and pine nut "risotto" (why the quotes?!). The hamachi collar with tare and shishitos, and the octopus that's somehow glazed in orange and chorizo, which I'm pretty sure is Mexican sausage (but what's it doing with the octopus?), and the sweetbreads, which I know are glands of some animal but I forget which—all I can really remember about them is that Jorge told us to tell people they "taste like fucking McNuggets"—and the pork jowl (how am I going to describe that without gesturing toward my face like Jorge did the other day?), and then, finally, the simple trout. Well, simple if it didn't have parsnips served five different ways with it.

Screwed, screwed, screwed is all I can think as the first diners are taking their seats.

Sofia stops me in the middle of the floor. Her eyes are all black pupils.

"You okay, Silver Fox?" she says reaching for my arm but missing it entirely. "Need a little something-something to take it down a notch?"

I will say, she does seem much calmer than I am, but I don't think I can afford to be anywhere near as calm as she is right now. If I know anything, it's that I don't need to try her flavor of vitamins for the first time in my life just now.

"No thanks, Foxy Fox," I say, as my first table is seated. "Wish me luck."

"Hey," she says, finally connecting with my forearm. There's something suddenly maternal and reassuring in her voice. "You've got this. After all," she says, "nobody's got more than three tables and the whole freaking staff is on. We literally outnumber the guests."

"Oh yeah," I say, almost calming down a bit. Even if things go really badly, I've only got three tables. How bad could it possibly get, I wonder, my feet still feeling wet and wrinkly after having mopped the floor at the other place for five hours.

Ordinarily—or at least as ordinarily as I've observed at high-end restaurants in the past—a server at a fine dining place won't have more than four or five tables at a time and will usually have lots of support to help take care of their section. Because this is our soft open—and not the knuckleheaded other place's—we've got eight servers and two food runners and two more wait assists for eight two-tops, twelve seats at the chef's counter, four round four-tops, and eight more configurable tables at the banquette, which can seat anywhere from eight deuces to four four-tops to two eight-tops. Fifty-eight diners max at any one time spread out over us eight servers. If my math is right, none of us should have to worry about more than 7.25 diners at any one given time tonight.

So Sofia's reassurance wasn't mathematically accurate, but I feel like she's more or less right. But, if there's one thing the god of mischief loves, it's confidence in a plan.

Already as I approach my first table—a two-top on the banquette that includes one of the managers from downstairs, the one named like he's an employee at a shoe store or an actual cobbler's son, Danner or Tanner or Cordwainer or something—I can tell that we're about to get overrun. Instead of seating folks in a staggered fashion like the best restaurants do—you seat in one server's section, then another's, then another's, and so forth, until the first server is ready to take another table—the hosts are filling the place up in each server's section all at once before moving on to the next section.

So already Rae and Rikki are full of deuces against the patio window, then Sofia and me in the round four-tops and along the banquette, then Cate and Claire's section mirroring ours, and then Esme and Hans finally along the whole chef's counter, and, literally, before it's even five after five, we're completely full. Not a single order has come out of the kitchen—I don't think there's even been time for an order to have been taken, never mind put into the Micros. What's going to happen is as inevitable as it is tragic: everything is going to shatter.

I thought this was going to be nothing like the other place. Instead it's beginning to feel exactly the same, but actually worse because I thought we knew what we were doing. I thought there was no way Dan or Linda or even Jorge would let this happen to us.

I do the only thing I can, which is to greet my first table as though nothing especially interesting were going on. It's more nerve-racking than I thought because I didn't really expect to even vaguely know anybody I'd be waiting on.

But that's not true.

I live in absolute sock-soaking terror of waiting on people I know, especially my students or my colleagues from school. But those nerves are quelled for the time being, for better and worse, because I don't really know Tanner and his girlfriend/fiancée/wife. Which is to say I know him because he's one of the three managers downstairs, but I doubt very much he knows me.

All I really know about him is that he wears these tortoise-shell-rimmed glasses that make him look bookish, and along with his Yankees cap he seems as though he's in witness protection—that and the potentially disturbing fact that his last name is the same as that one batshit-crazy state representative from a suburban district who clearly regards Sarah Palin as her sole role model. (I'm pretty sure they're not related, but I'd change my name if I were him.) And I can already feel the sweat trickling down my ribs.

"Hey," I say, stretching the word out because I need to do something to alleviate the facts that (1) I know who he is but know he doesn't know who I am; (2) I feel morally obligated to not say "guys" or "you guys" to a mixed-gender party; and (3) I feel similarly conflicted about calling anyone "folks" who aren't over fifty or so. "How are," I say, "you . . . two?"

Tanner almost but doesn't quite look up. His GF/F/W doesn't even lift her gaze from her phone, which she stared at even while they were guided to their table as though her device were a Geiger counter and she's prospecting in search of the mother lode of P-32 uranium.

"Sup," he says. He looks at the front of the menu as though it were a picture he doesn't like and quickly flips it over to see if there's something more pleasing on the back.

"So," I say, by way of throat clearing. "We've got a great menu for you tonight," I continue, already feeling stupid—about to explain our menu to someone who has worked in this building many more months than I have—but I need to start somewhere. "As you can see, it's divided up into four categories—"

"Great," Tanner says, but he means "stop talking," dismissive as though he's replying to somebody trying to sell him used tires. "I think we'll just go with—what did we say, babe?" He poses this as a question but doesn't actually look at his GF/F/W. "We'll split the salad, and then we'll get the chicken." He pauses for a moment, again flipping the menu over as though something new might have appeared in the intervening ten seconds since last he looked. "Yeah, and we'll split the chicken too."

I'm flummoxed for several reasons and have to again make a list to myself. One, it's my first table in about ten or so years. Two, he's a boss downstairs—in, it might need to be said, a *restaurant*—and so it logically follows that he knows—oh, I don't know—how to interact with servers, how to read a menu, how to order food, and so forth. Three, I don't know what Chef's policies are toward splitting things, but I doubt it's genial. Four, the two things he ordered aren't on the menu.

I flip open my server's book with my notes and the little crib sheet with nearly all the menu ingredients for every dish that I copied from Rae, but I know the words "salad" and "chicken" are not on the menu. I start with my first hunch, remembering the beets.

"Super," I say, "so then did you mean the beets or the asparagus, or was there something else, because, you know, it's our first day and all and we don't really have . . ." I trail off like a simpering fool so I don't have to call them out.

"The asparagus salad," Tanner says. This time he looks up at me, fully prepared now to treat me like one of his downstairs servers he just caught using a cell phone on the floor or putting my server's book down the back of my pants or some other cardinal server's sin.

"Right," I say, writing it down in my book, furling my forehead in faux concentration as I write out the entire word ASPARAGUS. "And then for your entrée," I say, even though we're not supposed to use that word—I don't even remember why. It had something to do with the idea that we don't have "entrées" or "appetizers" but rather small plates to share and some larger plates to share—kind of like Spanish tapas, but Jorge doesn't want us to talk about Spanish food because that was the last restaurant he ran and what we're doing is not that (despite the fact that there's a bunch of Spanish and Mediterranean food on the menu and all of our dishes are the size of tapas).

"So did you mean—?" I try again.

"The chicken," Tanner says, supremely annoyed now. "Or whatever Jorge's calling it—the hen."

I've so long sworn myself off describing anything that is not chicken as tasting "like chicken" after reading Bourdain's book that I had completely and honestly forgotten that a chicken is basically exactly what a guinea hen is.

"Awesome," I say. "Got it." I make a beeline for an open computer while it's there to be had.

My other two tables are already there and waiting, but I'm grateful beyond words that one is a couple of guys seated at a four-top so, at least for now, things aren't as bad as they can get. That said, the other couple looks kind of like Tanner & Co., already ready to be bored and/or irritated. I debate greeting them first but decide I better buckle down and see about getting this order in.

Micros, as a system, is laid out in a very symbolic and easy-to-use format. Management can change things around so that the main screen displays the tables more or less as they appear on the floor—which is great and super useful, except for the fact that the room is a box on the screen with no way to figure out geographically which way is north/south or east/west. If you were standing where I was standing, you'd figure—or at least I did—that, as I look at the screen, the top is the far wall and the bottom is the near wall and that left was left and right was right.

But that's not the case at all.

The north wall is the kitchen (which is not depicted on the screen) in real life, but on the screen, the north wall is on the left side of the screen. The top of the screen is the east wall, the bottom the west, the right the south. That's all well and good if you're following, but I wasn't at all. Basically, I was as confused as somebody trying to simply go north but the map I was using was oriented to the east like some side-scrolling video game. And also, I was facing east. I might just be dense, but to endeavor to share my struggle, next time you need to use a map, turn it ninety degrees to the right and see how that goes. It also might help to have a suddenly full restaurant of people who want to eat now and a string of servers lining up behind you who can't do their job until you get off the damned computer.

Suffice it to say, I decide to go throw down a couple of small plates for Tanner and his date to share their food when it arrives in lieu of walking the plank of asking Jorge about splitting food right now.

As I scuttle over to my next table, I realize that I failed to ask Tanner what he and his date want to drink.

Fuck it, I think, for the time being. If there's anybody in the building who knows how to get what he wants to drink, it's Tanner.

I decide en route that I'm going to have a better second table than my first one. I will be stronger, more articulate, more confident, more . . . I run out of time to really find what it is I want to be by the time I get there.

"Hey, guys," I say, somewhat linguistically relieved that they're two dudes, but my appearance at their table does not alleviate their gloom. "How are you tonight?"

Boy, oh boy, am I on a roll now! I think. Not only have I properly greeted them according to their gender, but I've also invited them to participate in some potentially witty and stimulating repartee, perhaps reflecting on the challenges and merits of the day's offerings and maybe even leading into some new and revelatory life changes for us all!

"Hey," the bigger of the two guys says. He looks at me once in what I'm afraid is an ungenerous appraisal. He's a meaty but still somehow compact guy, and he keeps leaning back against the banquette as though he is trying to will it to become his living room La-Z-Boy and is irritated with me for stopping that from happening. The guy across from him looks as wiry, feral, and mean as a raccoon who's jonesing for his methadone. He doesn't look at me at all and instead stares straight ahead as though trying to see through his compadre's face.

"Tell Jorge to send out whatever," the first guy says, and he gives the menu a dismissive twirl toward the center of the table. The second guy says nothing but twitches and then, with one finger, shoves his own menu forward without further elaboration, tattoos everywhere on his hand.

But I don't stare as I pick up the menus, somewhat relieved at the fact that I don't have to write anything down, and then remember to ask about drinks.

"And what can I get you gentlemen to—"

"Just water for him," the big guy says, "and I'll have the lager."

"Great," I say, "and we offer that in a four-ounce tasting size, an eight-ounce—"

"Just bring me a pint of the lager, whatever the hell it's called."

His face is beginning to tremble with irritation, but I can't help myself. "That's funny you should say that, because actually our lager is called 'Hell,'" I say, eager to prove that I know one bright and shiny thing about this whole endeavor. "'Helles' is the German word for—"

"Great," the guy says, his jowls beginning to tremble. "Super. Just the beer and some water for my friend."

Though it's a long way away from the firebombing of Dresden or any other of the various atrocities Kurt Vonnegut endured, I can't help thinking, "And so it goes."

I turn around and get a drink order from Tanner's table, and when I turn around again to greet my four-top, they're gone. I run over to the hosts to see what I did wrong, but they just shrug, as things are apparently going no better for them. They seem to be having the same orientation problem with their iPad that I'm having with Micros, so I leave them to their own little quagmire.

Now that everyone who was waiting has been seated, I realize I probably have a good while until my third table will be filled, so I try to relax as I approach Jorge on the line to tell him about my second table. Bella, I notice, is gone, and Jorge himself is doing expo.

I also realize as I move down the line around food runners pulling plates off of the stainless steel counter that this will be the first time I've ever talked to Jorge directly.

"Chef," I say, gulping, "may I call?" I hate how simpering the expression sounds coming from my mouth, but I know it's the formal way to approach a chef and am certain I have earned nothing more casual than that.

Nonetheless, he doesn't say anything.

He's reading all of the tickets stuck to the blue painter's tape on the counter. Dustin passes a snack board to him, and Jorge takes his grease pencil from behind his ear and crosses it out on one of the white tickets.

"Hands!" he yells without looking up.

I squirm and he shoves the snack board to me—a precarious dish that is actually a piece of what looks like driftwood about six inches wide by almost two feet long. Because it isn't a plate and has no lip and is so awkwardly long, with all of the fancy snacks placed so deliberately and artistically, it feels like a catastrophe waiting to happen.

"Table 202," I think he says, roughly shoving the board in my hands—still without looking.

"Two-oh-two? That's what I was coming to talk with you about," I say, a little shocked and entirely unsure of myself. Was that the table I meant? But I hadn't talked to him about it yet. Had I? There's no way I had. Did Sofia slip something in my coffee? Oh fuck oh fuck oh fuck.

"Table 202? What?" Jorge says. He looks up from the deep interior of his headspace like Kurtz ready to put another head on a pike or whatever it is he needs to do to create some order in this place.

"That's what this snack board is for?" I stammer. Dustin looks up from the other side of the pass and—I think—shakes his head at my incompetence. "They said to send out—"

"I know those assholes," Jorge says without looking at them or me again. "I'll send them a bunch of shit. Take them the board before the Reuben turns into a fucking hockey puck."

I almost say, "Oui, Chef!" as one former boss had insisted, but I trust my instinct that saying that to Jorge—at least in this moment—would lead directly to my face getting intimate with his fist, and I turn tail and run the board to my guys at 202.

In my gut, I know I am supposed to do what Cate did when she waited on Jenae and me the first time we ate downstairs and ordered the charcuterie board. Patient and kind, she described everything on the plate. The whole beer hall was complete pandemonium—not some metaphor but like the real place for all of Milton's busy, lusty demons in *Paradise Lost*—but Cate was unflappable and knew it all but said everything in a hypnotizing and warm way. Every meat, each cheese, all of the nuts, jams and jellies and mustards—even the board itself she described. "It's recycled!" she announced proudly. "Wood from the Hood is the company. Supercool, right?!"

But not me, not now.

"And here," I say, setting the board on the table between the guys at 202 like it's a self-explanatory rabbit I just pulled out of my hat, "here is the snack board. Enjoy!" And I skitter away to get the one guy's beer before they can ask any questions.

Tanner and his date already have their asparagus and are pecking at it while both staring at their phones. In that way they resemble a prison inmate and his visitor, separated not by a pane of vertical glass but rather by two little glowing horizontal ones in their palms. I ask how everything is so far, and they both say, automatically, "good," and I don't push it.

I head to the nearest side station to gauge the fullness of the water pitchers (both full) and check to see if we need more printer paper (of course we don't) and then check my server book to see if it has anything meaningful to distract me (it doesn't—I can't study any more right now and my first table has everything they ordered so far and my second table didn't technically order anything other than the one beer), but I've got to do whatever I can to stay away from the line or Jorge is going to make me run more food that I can't identify or describe.

"Hey, Silver Fox," Sofia says, giving me a hip check as she chicanes her way into the station and onto the open computer. "How's 202?"

"Who?" I say, barely registering the fact that she's talking about one of the only two tables I have at the moment.

"Your bros second from the end of the banquette," she says. She glances at them and me to make sure I've made the connection and then starts vigorously punching stuff into the computer like any of my millennial students, having used technology since birth.

"I don't know," I say, not yet sure how dishy we can be about our current guests. "They're okay, I guess."

"They're total dicks," she says. "The guy at position 1 is on the line at a pseudo-fancy place in Minneapolis that's actually just an upscale Arthur Treacher's," she says, popping her fingertip against the screen as though to articulate her comments. "And the guy in position 2 was hot shit a few years ago—on the short list for a Beard award and all that—but then started fucking his pastry chef and then her replacement and then anything with a hole on the front-of-house staff and then the booze and the cocaine and, blah blah blah. It's sad but not very interesting. I guess he's gotten clean and is trying to make his way back into the industry."

Like a couple of off-duty coroners, they're sampling the snack board in a kind of appraising way with intent and judgment but without any conspicuous enthusiasm, never mind enjoyment.

I guess I've never really met guys like them, other than that other Beard award nominee who tried to whip the lake place into shape before it opened. Guys operating at the pinnacle of achievement in the culinary field. Guys—and gals for sure, too, but I just haven't met them in my paltry experience yet—who are so far beyond the realm where a steak and a baked potato are anything worth discussing. Chefs who try so hard to break the binds of conventional ways of thinking about food that in some ways it kind of wrecks their ability to enjoy it anymore, regardless of the level of its conception, sourcing, preparation, or execution. My hunch is that it's kind of like expecting LeBron to be stoked to go hang out and watch the Bucks hack it out with the Timberwolves when neither of them is in playoff contention. Which is to say, sure, of course he cares about the game and is going to study his competition to see if he can steal any moves or better prepare himself for the next time he plays them, but is it even possible for him to enjoy basketball anymore as a spectator? What would it take for somebody to surprise LeBron anymore?

Not that these dudes at 202 or Tanner and his gal at 201 are exactly the LeBrons of anything, but when restaurants have become your entire professional world, a little of the polish and sparkle has got to dim.

Despite my initial hiccups, except for my one open table, the restaurant is completely full and nothing tragic has happened yet as far as I can tell. The bar is slammed and so is the kitchen, but both seem prepared for such an ordeal and, accordingly, no one is so far conspicuously dying or even bleeding out. Neither does anyone, I'm afraid to say, appear to be having that great of a time.

I recognize several faces from the staff downstairs, and many other tables seem partnered not like dates but like judges and juries. The difference from the other place's soft open is apparent finally. Their soft open was meant to try to figure out how many and various the ways were in which the restaurant will fail—not conceptually but practically. Let's see, their soft open seemed to say, how much water this jug with all these bullet holes will hold.

Here everybody has got their shit adequately dialed in so that the practice of making this work is beyond question. Here it's the idea— the very concept of what a restaurant can be—that they're trying to work out. They don't want to know if it can work but rather what kind of sense it makes. Or, more likely, what kind of statement it makes.

The difference is worth elaborating on, I think. The lake place endeavored to be simply a basic restaurant that served vaguely southern food. Rare would be the guest who showed up and said, "Chicken-fried steak! What in tarnation is that?" Rather, they were going to look at the menu and decide if they wanted it *here* rather than someplace else.

The Brewer's Table, apparently, feels like we're trying to figure out not what kind of cuisine we want to serve but rather *What is cuisine?*

Tanner & Co. demonstrated to me in spades the risk of posing such a question because, simply put, not everybody cares. Some folks just want their tacos—or whatever—their salad, their chicken. They don't want to think, But what *is* a taco? Is a taco still a taco if it isn't served on a tortilla? Is a salad still a salad if there's no salad?

Knowing now the conflicting things I did about my second table,

I approached them after they were about halfway through their next round of plates that Jorge had sent. I had hoped for a softball or two that I might by some miracle know the answer to.

"On the asparagus," the twitchy guy says. "How long does the egg soak in the tea? I just don't see how this much color could possibly have come just from tea. I mean, what kind of tea is blue anyway? What kind of tea is it?"

I barely remembered that there was a soft-boiled egg served with the asparagus and, now that they mentioned it, yes, it was soaked in a kind of tea, but I knew neither what kind nor how long—nor, for that matter, why anybody would soak any food in tea at all. It did, however, look really cool, like a soft-boiled egg that, with the shell mostly shattered but still intact, had been dipped and dyed like an Easter egg, and then, when the shell was removed, the dye had seeped through, but only where the cracks were, so that it took on this crazed and spiderwebbed kind of design but was entirely and perfectly edible. And, yes, a little blue.

"What a great question!" I say, as motley and ignorant a fool as could be.

"It can't be tea," the big jowly guy says. "I'm not getting any tea on my palate, and anyway it would probably take days to soak through the shell and inner membrane, even if it was partially cracked, and then there's no way you'd get a soft boil—the yolk would be hard as a rock."

I'm relieved now only by the fact that they clearly expect nothing by way of a reply from me; they're simply talking to each other and I happen to be there for it.

"The food runner said there was chorizo in the octo," Twitch says, again not looking up, gesturing to the partly eaten octopus Stonehenge between them. "Did you get any of that?"

"No," Jowly says, "the citrus pretty much plowed right through that, but not in a bad way. And hey," he says, now looking up at me, genuinely, it suddenly seems. He moves the octopus over and slides another square white, almost empty plate toward me so I can get a better look at what's left on it. There's something golden-looking and crumbly on one side, and on the other, bright oily swooshes here and there. Then there's a stem of what I now know is micro cilantro and a couple of little seed/spice/pepper thingies here and there, but that's it for clues. "So on the crudo," he says, throwing me the smallest of bones. "What," he says, pausing in anticipation of my failure, "what is that dot?"

I feel suddenly feverish and clammy and my palms start to sweat and I can feel my pulse in my neck and I think this is it: he's found a hair or a bug or a piece of metal or plastic or Lord knows what on the plate and wants me to answer for all of my and Chef's and the whole restaurant's transgressions and surely within a moment, or at the longest a minute, my fate will be decided and it will be spelled in bones and blood. But no.

"I can't quite place it," he says, not in the least bit cranky about it. He's musing, instead, intrigued by the fact that he doesn't know what exactly he's tasting and that neither does his esteemed, albeit strung-out, friend. "It's earthy, funky, dirt-like . . . almost like a mushroom . . . definitely a fungi . . . tons of umami . . . but what? It's too early for chanterelles and it can't be . . . it's not . . . is it?"

A broad smile breaks across his chunky, pork butt of a face. He looks at his friend. He looks at me. He looks at Jorge, but his back is to us so he looks back to me.

"It's huitlacoche, isn't it?" he says. LeBron, bemused. "That crazy motherfucker."

"You got it!" I say, and in this fashion, I suppose, my job for the day is done.

Chapter Eight

IMAGINARY BEER

THE REST OF THE SOFT OPEN ROLLED ON WITHOUT ANY FUR-
ther trauma. There was really only that one big push, and then we were
done and things wrapped up not long after nine. It felt initially like the
End of All Things, but it could've been way worse. I'm very concerned
with the pacing and overall willingness on the part of our hosts and Dan
to slam the entire restaurant all at once, but maybe the sections will stay
small enough where that won't be too terribly violent. I try to imagine it
like I'll be working at one of those crazy busy restaurants at an airport.
Not the special ones you find from time to time but the regular places
where business travelers and families alike seek comfort, familiarity,
speed, and consistency. And also room for their stupid roll-aboard suit-
cases. You know the ones. Restaurants that are unreasonably busy no
matter what time of day you encounter them. Monday at nine in the
morning? Packed. Sunday night at eight? Slammed. Wednesday, all
day? Wall to wall. Places where every hour of every day is happy hour.

What must it be like to work in those places? They've got to make
absolute bank. Their sections seem to be around five or six tables per
server, and their sections are full when they clock in, and they're still
full when they clock out.

That's what I try to imagine for myself at The Brewer's Table. So I'll
have four four-tops max, maybe six tables if they're deuces. That's as
bad as it can possibly get. I hope.

It's not like some of the restaurants I've worked in where you don't
have a section proper but instead you and a couple other servers have

to alternate tables whenever somebody comes in the door and you've got the main dining room, and then the bar, and then that damned back dining room that management will open up on the slightest whim because so long as they can keep putting butts in chairs they will and the servers can either keep up or quit (but not until after they suck it up and finish the shift). So instead of being able to have tunnel vision on your section, you've got to have this crazily evolving map going in your head and/or in your server's book where you remember you've got that couple by the door working on their appetizers, the four-top by the bar who all ordered different courses, the couple with their kid with the strawberry allergy in the back by the kitchen door, the group of six guys who just sat down by the TV and want me to turn the channel to the Packers game and also to have one of the female servers wait on them, and then there's the new family of five waiting, looming at the door, and you're pretty sure it's your turn again if you can get Lydia or Mandy to take the Packers table but neither are in sight at the moment and you accidentally make eye contact with the guy at the door—and on and on it can keep spiraling until you collapse or the clock brings the only mercy you'll know on this day, its end.

Incidentally, next time you're in a restaurant and it seems like you can't get your server's attention, take a quick tally of how many tables she has to take care of. Also whether anybody else is running her drinks and/or food. Lots of restaurants leave servers to fend entirely for themselves in way-too-big sections with no support staff other than some crack-addled assistant manager who waddles around every forty-five minutes with a gelid carafe of decaf pouring refills. If that server—if *any* server—has more than six tables (provided they're more than just two-tops), she deserves not only your mercy but the mercy of any and all available deities because she is being ground to little bits one Taco Tuesday Two-for-One Margaritas at a time.

On the other hand, if you can take in the whole dining room and your server isn't talking to more than two or three tables and isn't running any food or drinks and is camped out gawking at the bartender's eyelashes for whole spells at a time—yes, well, he might not be a pro and you might just leave a twenty for your drinks and just ghost. Because in the dining room—kind of like Bourdain's dictum about places with dirty bathrooms—whatever you see in the dining room is not only what any given restaurant is permitting to go on but what they *want* to go on.

At most establishments—whether at a high, low, or in-between restaurant, diner, falafel hut, tiki bar, pedal pub, sports bar, or popsicle stand—most servers are working their asses off whether you can tell or not. Like those with mandatory military service, countries that required mandatory restaurant work for at least a year would make everyone more compassionate, understanding, and gracious. Because, not unlike serving your country in a time of war, it is not possible to have been a server or a cook and not known trauma.

Most restaurants are the equivalent of an improv dinner theater, where the waitstaff are both the actors and the servers and they have no idea what direction the play will take at any given moment, but regardless of how comic or tragic the next turn will be, their job is to be calm and collected, knowledgeable, friendly, considerate, interested, invested, and unflappable. All while doing their damnedest to keep you happy and healthy without concerning you that any of the above is hard or dangerous in any way.

I keep forgetting, even on a linguistic level, that the "they" I speak of, yet again, is me.

After the soft open is over, we polish silverware and glassware, fold fifty napkins each, and tidy up the restaurant for tomorrow's first real shift, decompressing. Basically, we've realized, tonight we were full of restaurant people. I didn't know this until Sofia and others pointed out who everybody was, but with that realization came the understanding that we weren't necessarily failing. Simply put, we weren't going to succeed no matter what we did.

A soft open at a real restaurant is like a dress rehearsal for a bunch of lighting and sound techs. They're not there to have a good time. They're there to check the levels, as it were. You don't fill a restaurant with your peers and competitors and expect people to be blown away. You bring them in so they can test out the handful of dishes you've been slaving over for the past few months and say something like, "Yeah, that ought to work." There are no standing ovations at soft opens.

I thought Jorge or Dustin or even Dan would want our thoughts and feedback, but not so much. We worked out a few kinks with the computer system, but nothing especially important was discovered. These guys had done their homework and many of them had opened several restaurants before. Today, nobody at Surly was expecting surprises.

Which, of course, was a big surprise to me and, in the end, the complete opposite of the other place's soft open.

After the work is done, almost everybody sidles up to the bar at the end of the evening for our shifties. The chairs are tall and comfortable once you're in them, but they are heavy and hard to move around in, and because the bar is one long, ten-seat affair, you can basically talk to the person on your immediate left or right but nearly no one else beyond them. If the person next to you is talking to the person next to her, unless she leans way back or scoots out to let you in, you are out. And, by the time I sit down, I'm on one end with only Charles, one of our food runners, to keep company.

Charles is perched on his stool like I imagine monks or owls do when they find their way into, say, a high-end restaurant that only served beer. Which is to say, pensive, thoughtful, content but sheepishly aware of being a bit out of place. It turned out that he was our only food runner tonight as Nikki, our teen Sinead, didn't show up and that a no call/no show here, as in most restaurants, is as good as quitting *and* being fired.

"Charles," I say, hoisting my glass toward him in a vaguely celebratory way, "You killed it tonight, man." Now that I think about it, Charles did such a good job that I hadn't even noticed Nikki wasn't there until we were almost done with the shift.

"Yes, well," he says, and we clink glasses. "I thank you. And you also."

Charles's mannerisms are cagey and chicken-like, but in a good way. I recall from our introduction icebreakers during training that he was—briefly—a middle school science teacher in one of the Dakotas and is now back in school so he can one day work full-time in a lab and never set foot in a classroom again. That all seems to lead in one direction, but all night I noticed that any time any kind of R&B, funk, hip-hop, or, especially, doo-wop oldie would come on the stereo, he would, mid-service, add an electric slide to his step. Or a boogaloo. Or a twist. Or a grind. Or a bump. Imagine Higgins from *Magnum PI* or Felix from *The Odd Couple*—either one will do—imagine one of them scuttling apologetically across their hardwood floor when suddenly James Brown comes on the stereo and, without warning, he kicks a leg into the air, spins, does the splits and jumps back up (and kisses himself).

Charles did just about all of that all night long and still managed to never drop a plate. He was in constant motion but never seemed to break a sweat. Suffice it to say, we all already love Charles.

"What did you think of the night?" I say. I'm still a little baffled that there was no post-shift dissection of the service and that not even anybody hanging out at the bar now seems to have much to say, but then again the rest of the staff are mostly on their phones, making plans for tonight (it's only a little after ten) or catching up on whatever streams or snaps or posts or whatnots I have no idea about have been going on online while they all have been at work and have been observing the strict no-cell-phones rule we didn't have to get screamed at to understand. I, however, can't just sit here pretending I've missed a bunch of urgent posts and text messages because unless there's something from my wife, there's almost certainly nothing at all.

"It was good, good," Charles says, nodding. "What about you? Good night?"

"It was okay," I say. "But to tell you the truth, I feel really out of my depth here."

"Yeah?" Charles says, his head pivoting on his bony shoulders like C-3PO's. "How so?"

"I feel like I don't really have a handle on the menu yet—I've got maybe 50 percent of any given dish down—and most of the beers are just names and hop profiles with nothing meaningful attached to them."

"OhmyGodIknowexactlywhatyoumean!" Charles says, his robotic posture suddenly collapsing with relief as though some kind of exoskeleton had been suddenly removed. "And Chef scares the absolute shit out of me. After he made Bella disapparate I was scared to get within ten feet of him, and he kept giving me this withering look like What the fuck is the matter with you? How do you expect me to hand you plates from ten feet away?"

I shake my head, but in agreement. "I know. Once he took over I started pretend-stocking the side stations with silverware so that I wouldn't have to run food. Sorry if that made your night rough."

"No," he said, grooving to the Stevie Wonder that just started playing. "It's all good. If I didn't have to do it all kind of by myself tonight, I probably would've puked during tomorrow's real service. Then again, the statute against puking in two consecutive days has not yet been ratified."

"Cheers to that," I say, and we clink our glasses again.

Charles takes a sip and I can tell he's probably about ready to pack it in, but I've got one more question for him.

"I see you're digging on the Stevie Wonder and were into the music more or less all night."

"Yeah," he says. "Can't get enough."

"Did you recommend any songs?" I say. "For the playlist or whatever Dan was going to make?"

He arches an eyebrow without turning directly to me, testing something out. "Yeah . . ."

"I recommended a whole bunch—put way too much thought—or at least time—into it," I say, "and I didn't hear a single one of my songs."

Charles shakes his head vigorously as though I had just confessed that I was thinking about quitting or starting up a new prescription drug habit. "I don't know," he says, finally turning completely toward me—which, given his litheness and the weight of the chairs, is awkward and takes some doing. "I had a few of my requests on tonight, but I just picked the no-brainers—Al Green, Funkadelic, a little Stevie," he says, nodding toward the speaker.

"Oh my God!" I say. "What the hell was going on with that one song? I think it was Marvin Gaye—it wasn't 'What's Going On' but something else about as popular—but out of nowhere—"

Charles purses his lips. "There was perhaps," he says, "a brass band involved?"

"Yes!" I say. "Like Marvin was getting it on with a middle school symphonic orchestra or something."

"It was 'Sexual Healing.'"

"Right!" I say, amazed Charles knows what I'm talking about. "A classic, but where the hell did the marching band come from? It was just too weird."

"Yep," Charles says, and suddenly I know where it came from.

"Oh God," I say, "I'm sorry. I didn't know it was your . . ."

"Mm-hmm," he says, holding his beer up to the light, watching all the little bubbles pop. "Bad news for you was that Dan didn't play any of your songs—yet. Bad news for me was that he did play a few of mine—and then stopped playing one of them, right in the middle of the shift when Jorge said, 'What the fuck is that shit?' and dropped what he was doing to go get Dan to immediately remove it from the playlist."

"I didn't . . . I'm sorry."

"It's a great fucking song," Charles says stridently, bringing his glass down to the bar a bit too forcefully so that I can feel a couple servers to my left turn our way. "They're called the Hot 8 Brass Band and they're from New Orleans and their all-wind-instrument cover of "Sexual Healing" is amazing—check out the video on YouTube. Even Quentin Tarantino commented on it—Quentin Tarantino!—and he said that he listens to that song every day. That song! Every fucking day! Quentin motherfucking Tarantino!"

"Whoa there, tiger," Miko says, peering out from the other side of the beer taps. "That first beer one beer too many?"

"OhmyGodI'msosorryI'msoembarrassed," Charles blurts, but Miko is clearly joking.

"Well, no wonder then," I say as Charles gathers his nerves. "Dan's never going to play my songs. My entire list was a bluegrass rendition of *OK Computer*."

"Oh my God," Charles says. "*Pickin' On Radiohead!* I love that album!"

Neither of us stays much longer, but I think in the end we both agree that it's been a mostly pretty good day.

After I get my kiddo on the bus the next morning, I know what I have to do. I've got almost eight solid hours before I need to be at work again, so I go for a short run, force myself to not send Dan or anybody else any more stupid emails about name and/or music suggestions, make sure my pants and shirt from last night are clean and pressed enough to wear again, and fold them back into my bag.

For the rest of the day, during any free moment—whether waiting for my dog to poop or in between bites of Honey Bunches of Oats—I try to concentrate on my beer knowledge. It's a weird thing to do at 8:07 a.m. on a workday, but I'm not drinking beer. I'm studying it.

By midmorning I finally have a pretty good little schema thing I came up with for understanding beer. Gary gave me the idea to put it on an axis, but he only plotted the specific beers we'd be serving, and since Surly doesn't supposedly brew "styles of beer" but only our own proprietary Surly beers, it was really difficult to understand things in a broader light. It was a long way from perfect, but for the first time it gave me a kind of visual perspective on why a certain beer might go with a certain food. For instance, let's say you're having something heavy, rich, and creamy. Would you really want something sweet, roasted, and malty to go with that? That would be like having a porter or an oatmeal brown ale with, well, oatmeal. Too matchy-matchy. Clearly, something as hearty and toasty as oatmeal needs to be brightened up with something less malty. Maybe a lager if you want something traditional (for

those of us who like simple beer and food pairings for breakfast) or perhaps a nice West Coast IPA (for those who prefer a little shot of citrus first thing in the morning and want a beer to act something like a mimosa).

Something was really beginning to click, and before I knew it, it was time to get on my bike and roll into work.

Downstairs, just before I start up the ramp to the employee entrance, one of the smokers with multiple face tattoos as well as multiple face piercings gives me the slightest (but nonetheless perceivable) "sup" nod. I suppose he might've just been gesturing his hair out of his eyes or thinking about the physics behind time travel, but I choose to believe otherwise and boldly return his (almost imperceptible) gesture.

Clearly, I am a made guy now. I'm in. I had made the cut. I was one of them. *Familia.*

And then once I make it past the scary ARC HAZARD box, I am right back to full anonymity as the beer hall back of house is fully abuzz with a slammed Friday crowd, so I opt to not be flashy about my full indoctrination into the tribe and instead bare it with subtle humility in keeping with the code of my not-so-new-but-new-here identity.

I go to change in the men's bathroom, and a couple of guys are speaking Spanish to each other and don't pause for my sake. They appear to be changing out of a different restaurant's uniform and into our own. Given the fact that I squandered many hours of my morning and afternoon just eating cereal and thinking about beer and that these guys had already been at work for the whole day and are likely just getting started on another entire shift, I feel obligated to say something—anything—to let them know that even if I'm not working the same exact way they are, I am still their brother. *Viva la familia!* I want to say.

That or something against the man like one of about four phrases in broken Spanish I know, such as *too mucho pinche trabajo, no?!*

But as I jockey around them and their conversation as I get into my work clothes, I say something not even I understand. "Is este es su lockér?"

The older, more sympathetic of the two looks at me in a wincing way. "What can I do for you?"

"Oh," I say, feeling more than a little self-conscious about the fact that I'm putting on a uniform only nine other people upstairs wear while these guys are donning the ubiquitous black check pants and short-sleeved white cook's shirt that about fifty guys downstairs wear. Much as I want us to be, we are decidedly not in the same family. "It was nada. I mean nothing. I mean I was just . . . is this locker? . . . are you? . . . is this your locker?"

He just shakes his head.

"Take your pick."

The elevator again takes forever, helping the embarrassment simmer, but I try to let the excitement and vast possibility of the big first night wash it away one vertical foot at a time, and I get out and remember Pat and all but scream "corner!" as I come in through the back swinging door, but nobody is in the prep kitchen or in dishland to hear me other than Christopher Walken's smiling paper face.

In the dining room nobody looks impressed to hear me shout "corner!" either, and everybody behind the line cooking and prepping is doing so like a bunch of novitiates saying their vespers. Miko, too, on the other side of the restaurant, is wiping down the bar with quiet concentration.

Rikki, Esme, Claire, and Sofia are proofing their sections already, and I check my watch, but no, I'm not late. It's just 4:20. We are supposed to clock in at 4:30, and pre-shift is supposed to happen just a bit before we open the doors at 5:00. I'm confused. I'm not late, so then they're early? All of them?

I decide to do the only thing that has consistently gotten me through this life—whether it was college, grad school, job talks with full professors and university deans, parenthood, my own class preparation, lesson plans, curriculum design, homebuying and repair, book writing, or, probably the hardest, restaurant job interviews with people who could smell shit from a country mile away: I just have to fake it.

So. I walk confidently across the dining room.

"What's up, Rikki? Hey, what's up, Esme? Hey, Foxy Fox," I say and then, for reasons no one understands, I add: "Meow."

Finally I make it to the server station by the bar, where I give Miko the same practiced nod that smoking server gave me downstairs.

"What's up?" he says. "You need something?"

"No, no," I say—did I really not nod right? "I was just saying—anyway—no. Hi. Yeah."

And I go to clock into Micros, but before I can tap the screen awake, Dan surprises me from behind.

"Hey," he says, "good night last night."

"Hey!" I say, trying to figure out if it was a question or not. Needing to pick, I just go with an opaque but enthusiastic "Yeah!"

"So, yeah, hey," Dan says. He looks down at the tops of his Chuck Taylors and then back up at me and then kind of scans the room to make sure he's not missing something. "What're you doing here?"

"I . . . well . . . it is opening night, is it not?" I say. I stare at the Micros screensaver and feel a kind of kinship with its eight-bit, old-school display of the word "Micros" bouncing around jerkily from one corner of the screen to the other.

"It is indeed," Dan says. He's not giving more than about half of his attention to me, and that is A-OK, especially considering the day it is and what his job is as opposed to mine. He's got some spreadsheet program open on his laptop next to the Micros, and its hundreds of needy cells I see is enough to send me backward off the balcony. It's hard to believe I even came up with my beer knowledge deal on an X/Y axis, so great is my fear of all things spreadsheet. "Anyway," Dan says, clicking a key with finality, "It's Friday. You're scheduled to work tomorrow."

"Ah," I say, but I mean it like a parry rather than an argument, "yes, but . . . it's opening night."

"True," Dan says, his grin a little less shit-eating just now. "But you're not scheduled."

"I just assumed that opening night was like the soft open."

"Except," Dan says, closing his laptop and tossing it as casually into his bag, "except that it isn't."

"Okay then," I say, a little chided, yes. A little hurt. But also still more than a little confused. Why wouldn't he want all hands on deck? Isn't it going to be more or less a bigger shitshow that yesterday promised to be? Aren't we just minutes away from him unlocking the doors to another horde of hungry and thirsty diners, this time no friends among them? Isn't it going to be Pure D nuts?

"Isn't it going to be Pure D nuts?" I decide to say out loud. My friend Bruce used to use that phrase, and, like most things Texan, I never really understood it but nonetheless often find myself using it.

"Purity what?" Dan says. He squints. I think he wants to like me, but I am really not giving him much to go on.

"Nuts," I say. "Busy."

"Actually," Dan says, waking up his phone to check the time, then looking around the dining room, then looking at his phone again. "You know what. I actually think we might have another no-show."

"Rae?" I say, recalling how Chef barked at her the other day and the fact that she had already bragged to me about how she liked to quit things such as other restaurants and the grad program at my school.

"Yep," Dan says. "But what we've got on the books isn't too bad, and I think we can run with four servers. Why don't you run food with Charles?"

"Deal," I say. "As long as there isn't any more marching band music."

Dan shakes his head, knowing, finally, exactly what I mean.

"Deal," he says. "But you've got to wear the short apron."

After Dan gets me set up in a support staff short apron—despite the fact that it is literally just a shorter version of the same black-and-gray broad-striped waiter's apron, it is a tad emasculating—I start to head over to the counter for pre-shift and bump into Sofia on the way. She's polishing the silverware at the same table she was on when I walked in about ten minutes ago. If it were a ten-top, that would be understandable. It is, I'm afraid, a two-top, with only four total pieces of silverware—fork, knife, fork, knife—as well as two very clean water glasses.

"Hey, Foxy," I say.

"Hey," she says. She's smiling broadly and her eyes are at least as wide as the rims of the water glasses as though she's just been announced as Miss Greco-America. "I'm not nervous. Are you nervous? I'm not nervous."

"Oh boy," I say, but she doesn't notice. "Are you okay? Too many vitamins?"

"Mm-hmm," she says, still wide-eyed and polishing. "Just one vita-
minny. It ought to kick in sooned. They."

"Yes indeed, I'm sure it/they will," I say. I feel like I ought to do
something or say something—maybe even tell Dan—but I know Sofia
is far more a pro than I am. If she needs a little something to get over
these first-day nerves, she probably knows better than I do. She seemed
a little off before yesterday too but ended up sailing smoothly through
the whole service. "You'll be," I say, trying to think of something inspiring but also reasonable, "foxy!"

"Foxy!" she says, holding high her one shiny fork, reminding me,
inexplicably, of Ozymandias.

I find my way over to Charles, who has been meticulously stacking
and aligning plates and bowls on the line between the kitchen and the
dining room.

"Hey, man," I say, "how's it—"

"Don't even try to horn in on my plate situation!" he says. His eyes
are crazed like Golem's as he shields the stack of plates he's been turning around and around since I arrived.

"You're coming in loud and clear, sir," I say. Even though it's the first
day, I already know the relief of having found a task to do so that nobody will suspect you of doing nothing. "So," I say, "turns out I'm not
supposed to work tonight, but Rae is a no-show so Dan wants me to
help run food."

"Oh," Charles says, dropping his defenses. "Cool."

Before we can get more in-depth than that, Dan's Europe track takes
over and "The Final Countdown" starts again blaring over the stereo.
He lets it play as everybody assembles on their respective sides of the
pass, and he continues to let it play well into the little-known second
verse when Jorge gives Dan that executioner's throat-slit gesture and
Dan rips his iPod out from the AUX port and hustles over to the line.

"All right!" Dan says, "our big first night! Let's get it started! Chef,
what you got?"

Jorge looks as pleasant and warm as a saguaro cactus.

"Not much," he says. "We don't have a ton on the books, but it's all
new, so give us a little more lead time than you usually would. That said,

pickup on everything other than the meat dishes shouldn't be more than five, maybe seven minutes. Even the meat shouldn't be more than thirteen, maybe sixteen minutes tops. And don't take temps on anything. We're not a fucking steakhouse—we don't even have a steak," he says, mostly to himself, smiling with the realization. "And no special preps or sauce-on-the-side bullshit, okay?"

Nobody says anything or even risks glancing around to see if anybody else has anything to say or ask. It is so clearly not the time for a long, drawn-out Q&A sesh, and we're all a little leery of him barking down our throats again or of him or Dan giving us another pop quiz.

"All right," he says, more soft-spoken, less asshole-ish. "We've been working toward this for—well you FOH guys have for a couple of weeks, but everybody in the kitchen has been at it for months and then some. Let's not fuck it up."

Dan raises a make-believe glass in his hand, and we all follow suit. "To not fucking it up!"

"To not fucking it up!" everybody cheers, and we clink our imaginary glasses full of imaginary beer.

And then, unceremoniously, we all split up to finish whatever prep stuff we've been doing. But before I can peel away to make sure Sofia is going to be okay, Charles pulls me aside toward the end of the line and glances over his shoulder to make sure Jorge is out of earshot.

"He's a mumbler."

"Who's a what?" I say. Sofia is back at table 41, reworking the polish on that same fork as though it's the lens for the Hadron supercollider.

"Chef," Charles says through gritted teeth. As though that will make him easier for me to understand and more difficult for unwanted parties to intercept his message. "Jorge! He mumbles."

"No," I say. I don't believe it. I mean, sure, he's said some things so far that I haven't loved as far as content and/or delivery goes, but he's the executive chef of this restaurant and the whole monstrosity below us. No way could you command a battalion like that without being able to, at the very least, enunciate. And, of course, he sounds okay in pre-shift meetings.

"You better freaking believe it," Charles says. "He starts out okay, but then when things pick up he garbles his table and position numbers like he's got a mouthful of cookies."

"No shit," I say, glancing at Jorge, his eyes flinty, his arms swarthy with tattoos.

"Yes shit," Charles says. "God save us both."

And then, before we know it, the doors open and our first guests have arrived—an older but not quite elderly couple that nobody appears to know. Esme goes to greet them with a wide smile and open arms, but they don't even look up and they shoo her away so they can look at the menu on their own.

And then there's nobody else for quite some time. At least fifteen minutes. Maybe twenty. And then another deuce comes in. And then nobody else. For quite some time.

After another hour, I haven't quite experienced the mumbling Charles warned me about, but he's pretty much been to Jorge's left with me right behind him, and every time there's been something to run so far all I've had to do is follow Charles.

Around seven a few larger parties finally begin to stack up around the host stand, and I'm afraid we're going to get clobbered, but Dan's there, checking the hosts' work and—what?—he invites three four-tops in a row to grab some drinks at the bar.

There are open tables, to be sure, and everyone can see them. But Dan is holding solid for some reason. I can't help myself. I need to know what's going on. I drop a snack board to 51, explain what I can, and go shoulder up to Dan, both our backs to the glass window overlooking the brewery.

"Hey," I say.

"Matt," he says. "What's up? Charles treating you all right?"

"What?" I say. "Oh, yeah, no. That's all good. I'm just wondering why you sent these groups to the bar. Last night it was like every table was sat at the same time. What gives?"

"What gives?" Dan says, cueing up his grin for some overtime. "What gives is I'm not a psychopath. What kind of manager would allow his hosts to fill an entire dining room all at once?"

"Um," I say, but I don't risk another word.

He pats me on the shoulder in a big brother way despite the fact that I'm probably at least five years older than he is. "Dude, yesterday was the soft open. Gotta see how far you can push the system before it breaks."

"So last night," I say, understanding several train cars behind the driving engine of Dan's words. "That was a test?"

"Yep."

"And tonight?"

"Yep," Dan says.

"The real thing?"

"And bingo was his name-o." And with that, Dan swoops toward the line to help Charles do the job I am supposed to be doing.

At 7:30 we are near to full, and Charles runs a two-top's apps, and just as he leaves the pass, Jorge yells "hands!" and he looks at the place where Charles had been and then, confused, finally looks my way and shrugs.

He hands me another snack board and gestures toward the beets that just came up at the garmo station down the line—it's short for *garde manger* and it's where they make the salads and desserts and other cold food so it doesn't get ruined by sitting near the hot food.

"All right," he says, clear as the tattoo on his arm that says "Don't burn the daylight away," which startles me, because I'm pretty sure that's from a Dave Matthews song, but then he continues, sotto voce, "table-iffy-un, sea- uh, sna- bo-, sea- -o, -eet."

Without meaning to, I give Jorge a wide-eyed look of terror and confusion. "What?" he says, and I can feel the pop and whoosh of the trash incinerator that is his heart flare hungrily so as to take care of me and all my idiotic bafflement.

"Nothing, Chef!" I say, and off I go, and as inconspicuously as one can, I do laps around the entire restaurant, checking in with each of the four servers until finally Esme puts me on the right track to her table.

Chef, indeed, is a mumbler.

As the night progresses, Charles gets bolder, and with every dish he runs, he returns to Jorge's side a little closer than before, and by the end of the evening, Charles has gotten close enough to be able to read the tickets over his shoulder, and all we have to do is wait for Jorge to yell "hands!" and he can stammer all he wants, but thanks to Charles we now know where the food needs to go before he tells us. This, it appears, is what Jorge wanted all along. To be strong and mostly silent as expo. Anybody can do that job and do it loudly and make it seem impossible and doomed. To do it with finesse and without a single word, I'm recognizing, is art.

By 8:30 or so it becomes evident that we already have served the last of our diners for the night, so Dan cuts me and three of the servers.

It feels a little like a letdown, given how long and dramatic the lake place's first couple of days went, as well as how at least that initial first push was during yesterday's soft open. At the same time, from a food delivery perspective, everything went as swimmingly as could be. I still have, of course, my own first real night waiting tables ahead of me, but having seen what our actual service will likely resemble, my nerves are put significantly at ease.

I have a quiet shifty at the bar by myself—the rest of the servers who were cut still had silverware to polish and napkins to fold—and I watch the bubbles make their languid way up through my Bender, a real beer to replace the imaginary ones we all had at the beginning of the shift. I try to savor it and call to mind all of the glory that I felt when Cate had described this beer as being like that childhood Christmas chocolate orange, barely able to handle all of the wonder and surprise that such a little container held. But now I'm so thirsty and tired from a fairly long day, and even though I could now map it on a X/Y chart, my drink just tastes like a real nice dark, brown beer.

Chapter Nine

WELCOME TO
THE OCCUPATION

I'VE WORKED IN MORE RESTAURANTS, AND FOR LONGER, I now realize, than I've worked in higher education. I suppose you could count my teaching while in grad school and my freeway flier years when I drove over an hour, each way, every day to teach at satellite branch campuses of Ohio State and the University of Wisconsin, but I don't. My job had more to do with long-haul trucking than it did education, and not just because I did so much driving. Given the fact that both my students and I also had part-time jobs on the side, just keeping everybody awake was the main goal, and no amount of caffeine, nicotine, or Ritalin ever really did the trick.

At any rate, I now keep my resume up to date and saved in my JOBS file because you never know when you're going to need it. I keep it right alongside documents like my annual faculty reports and my academic CV, which I update every year even if I haven't done anything remarkable other than flip twelve pages on the wall, as well as some long-shot, aspirational dream job letters to such ridiculous places as Carnegie Mellon, the University of Vermont, and Bowdoin College.

Also, I'll probably never get another job in higher education if I ever leave the one I'm at (or get fired).

The job market is ridiculously flooded with unemployed or underemployed but still super-qualified grad students who already have, like me, terminal degrees in their field and have published more than enough to earn a shot at a good teaching job. I was lucky and came of

teaching age when the field of creative nonfiction was new and bur-
geoning and almost every college in the country was fighting over one
another to hire only a few vaguely qualified applicants such as myself.
Now the market is more saturated with highly qualified applicants than
most professional sports are with up-and-coming talent. It turns out
you need not several hundred writers and teachers of writing to work
in higher education in any given city. Nope. In a metro area as big as the
Twin Cities, you need about twenty or so. And, since they don't tend
to die off as a pack every year but rather only in ones and sometimes
twos every several years, with the boom of MFA programs, the pools
of applicants continue to swell annually with each program turning out
many times the number of graduates relative to the open jobs.

So, realistically, if things don't work out at Surly or the lakeside place
and I still need to make more money, it's probably going to be at another restaurant, not a college.

Looking back on it, my trajectory in the restaurant world feels like a
more formal course of study than any I undertook to become a teacher.
I went to a ridiculous amount of school—undergrad, an MA program
in English, an MFA program in creative writing, and a PhD program
in creative writing and literature—but when it comes down to it, I can
count on one hand the number of classes I took on any given subject.
One intro to fiction class, one drama class, two or three grad nonfiction classes, maybe four grad fiction classes, maybe a couple more here
and there in my specialty area of creative nonfiction that I'm forgetting, but not many more. And so I wrote a manuscript that became my
dissertation that, many years later, became my first book, but so what?
That makes me a so-called expert in teaching and the art and craft of
creative writing?

Frankly, I think most good bookstore employees know more about
what's going on in creative writing in America than many who teach
in undergrad or even graduate programs. More than I do anyway. If
the *Jeopardy!* category was contemporary literature, I'd be caught flat-
footed in a contest against pretty much any of the fine folks at Next
Chapter Books, Subtext Books, or Magers & Quinn.

I'm not trying to dis any of my college colleagues. All I'm talking about is my experience, and the only reason I want to do so is to demonstrate this: working in a restaurant and becoming not just proficient but actually *good* at it is at least as meaningful, difficult, and rewarding, if not more so, as the pursuit of any other ostensibly meaningful career.

I have several friends and colleagues who simply don't believe this. "But why the hell does this matter to you?" they demand. "You're an author, a professor, somebody who seems to have your professional life in order—other than the mathematic fail of not taking into account the financial hit that would come with a yearlong sabbatical, you seem to have your ducks in their respective rows. Of all possible ways to make a little money on the side, why the hell would you want to wait tables?"

Two, maybe three reasons.

Reason 1: I can write at will. I cannot publish at will.

I might never get paid again for writing, no matter how many manuscripts I write. Sure, I sold my first book to one of the big New York publishers, but I've never seen a penny—literally—in royalties, and I've got three (four if you count this work in progress) manuscripts complete but lying dormant, slowly decomposing in my drawer.

When we're young and up-and-coming, we writers think there's nothing more difficult than publishing your first book, but I now believe that isn't true. When you're up-and-coming, you've got nothing but raw potential ahead of you, and if it's one thing the American public/publishing industry swoons over, it's the promise of a new, undiscovered talent. What's not to love about an emerging writer? Who knows how many thousands or hundreds of thousands or millions of books she'll sell?

I have it on good authority that after you've published your first book, when you go to show your agent or editor a second, or third, or fourth, manuscript, the first thing they do is pull up your previous sales on some dark web website called Book Scan where they can see exactly how many copies of your first book you sold. You no longer have the unbridled promise of the emerging writer. You have precisely X promise based on Y number of books sold in the past, and anybody who reads any of your subsequent manuscripts will calibrate her enthusiasm accordingly.

Reason 2: I can, apparently, still wait tables almost at will. And I get paid immediately.

No restaurant manager before a job interview looks up my sales record from my previous job. And usually they don't even want to know why I left any of my previous jobs. Turnover is just a fact of life in the restaurant industry. What seems to matter most is what you can do today.

Reason 3 (and I'm bound to get in some hot water here): I like teaching. I *love* waiting tables.

But wait, my dear colleagues/department chair/college dean/executive vice president! Let me go into a bit of backstory first while the university lawyers start poring over my contract, the faculty handbook, and tenure law.

For the most part, a classroom is a classroom is a classroom. For sure, not all of the students are alike, and every given class—even every given day in every given class—has the potential to be a winner or a stinker for reasons that often as not have nothing to do with the material being taught or the amount of passion or preparation I put into things. Everywhere across the continent, college students are sometimes bored and tired and sometimes perky and invigorating (thanks campus Adderall dealer!).

I enjoy teaching very much, but at the end of nearly every day, most of my students regard my classes as—and this is a direct quote from one of my student evaluations—a course that "I thought was going to suck but didn't."

Restaurants, on the other hand—if they're good ones—well, we don't just nourish. We delight.

Every job I've had in the restaurant world has prepared me in some significant way, I think, for what we're about to do at Surly. Even the jobs I didn't get have something to do with how I wound up here, so allow me to indulge in another litany.

The first job I applied for in a restaurant was an early favorite of mine that my mom took me to when she visited me in college. It was called John Hawks Pub, and it was nestled just above the Milwaukee River as it ran its course through downtown Milwaukee. It was dark and murky and smelled of stale popcorn, malt vinegar, and sour British ale. I loved everything about it. It was the kind of place where I imagined all of the great Romantic poets and British novelists ate, drank, wrote, and/or expired. I imagined it was where Wordsworth would brood over the first-person singular versus plural in one of his sonnets ("'The world is too much with me?' No, no. 'Too much with us?' Bingo!") or where Dickens would secretly embed himself amongst the urchins while the latest installment of *David Copperfield* was read out loud by some would-be actor/bartender. It was certainly among the last places Edgar Allan Poe or Dylan Thomas staggered out of before choking on their own respective vomit. It was one thing to be living the life of an English major among a bunch of twenty-year-olds who couldn't be bothered to change out of the sweatpants they slept in to come to class. It was quite another to feel like I could actually immerse myself in the milieu where I imagined much of our greatest literature came from.

Suffice it to say, with my restaurant work experience being precisely zero, I didn't get the job. I had thought that they might be impressed with my two years of experience in high school selling skateboard parts, but I was mistaken. They never even called me back.

Much as I want to say that I let that disappointment just wash down the drain, I was crushed. It didn't feel like I didn't get just this one job at this one restaurant. It felt like the *kind* of job I would never get. I imagined that working there would be somewhere between being an artist, an entertainer, an actor, and the host of an ongoing party. Like Billy Crystal hosting the Oscars—that's what the people who worked there seemed to embody. They were jolly, witty, winsome, and gracious. Even though I only went there with my mom, I felt like I was always smarter, more attractive, more popular, and, as much as I hate to say it, more welcome there than I was anywhere else.

Until I wasn't.

I never went back there again. And, like most restaurants, it too is no more.

It was on that site in the mid-nineteenth century that Milwaukee's first mayor, Solomon Juneau, printed the first issue of the city's once-great newspaper, the *Milwaukee Sentinel.*

It is on that site that now John Hawks Pub has become an august Gold's Gym, and from your treadmill, through its river view, you can meditate on the bronze statue of another Milwaukee icon, one with an even arguably greater impact on our town: Arthur Fonzarelli.

137

* * *

It took more than a little time to lick my wounds and try to find work in another restaurant. Between my sophomore and junior years, I went back to work selling bikes and skateboards and box after checkerboarded box of Vans shoes. I loved it when I began, but I had worked there on and off for four years by that point, and its hold on my life was slipping. The money was decent and my boss had started giving me a commission on bikes and even started making hints that I was potential manager material. He went so far as to help pay for my schoolbooks and to entrust me to drive his GMC pickup and the twenty-foot, gooseneck trailer full of BMX parts all the way out to the indoor bike-racing mecca in Walworth, Wisconsin. (Don't pretend you don't know of Walworth—it's where they make Kikkoman Soy Sauce, of course.) But not even selling sprockets and chains to the up-and-coming twelve-year-old BMX racers of greater south central Wisconsin could persuade me to stay for very much longer.

When I drove back to school, I happened to come across a HELP WANTED sign in one of the seemingly down-on-its-heels places just off I-94. It wasn't the Glocca Morra (or, as the rougher drinkers on Marquette's campus used to lovingly refer to it, the Glock) but rather a couple blocks farther west than that. Miss Katie's Diner.

Before I applied, I tried it out with my mom on one of her visits and was modestly impressed. To be sure, it was never going to compete with the antiquated, poetic vibe of the British place on the river—after all, the only river in view of Miss Katie's was the stream of traffic on the interstate—but it had a kind of quintessentially American, midcentury thing going on. There was no James Dean in his white T-shirt and leather jacket, but there was this guy the waitresses called, well, "Guy" because that was his name, and he was chain-smoking More cigarettes in the middle of the dining room, his only order, apparently, a small glass of milk. What's more American than that?

There were two rooms. The bar, where it appeared only men worked and drank, and then the dining room, where only women worked but all were otherwise welcome. My mom and I ate in the dining room, of course, and we were warmly greeted by the last women of the world who still kept weekly hairdressing appointments—women who wore nude stockings—women whose names have only barely limped into the twenty-first century. I can't recall them specifically now, but I'm sure there was a Phyllis, a Gloria, a Dolores, and probably even a Blanche. Whoever they were, they served us "chocolate malteds" and "steak casino"—the first was just a shake with a bunch of malt powder in it; the second, inexplicably, was just a hamburger without the bun—but it was all brilliant. It was Americana at its most authentic, and I wanted in.

After our meal, I asked Wanda if I could get a job application, and she just nodded her beehive toward the dark bar. Much as I wanted to bring my mother with me, I went alone into the red-neon-lit room and waited until the barman noticed me. (It was quite some time.)

For the record, he didn't even say "yeah?" but he did almost imperceptivity lean my way.

"Can I . . . may I pick up a . . . an . . . a job application?"

The man had hair that might have been put on his head in one piece—it might even have been his own hair—but it was so shellacked and unmoving it was difficult to imagine it as a piece of him anymore. He did this thing with his lips that made me think at first that he was practicing kissing and then that he was trying to get something out from between his teeth, and then he did make a sucking sound while he slowly wheeled his head and hair toward what was presumably a man behind a newspaper.

"Peter Jr.?" the barman asked.

"Yeah, Johnny?"

"Got any of them job applications around?"

The newspaper rustled, lowered—but not much—then rustled again.

"Tell him to come back at five."

The barman licked his teeth from behind his closed lips.

"Peter Jr. says that you should come back at five."

"Five," I said. "Got it."

A couple of hours later I return, and other than the fact that I am no longer with my mom and the quality of light has shifted to an early evening gauzy affair that slanted lazily through the south-facing picture windows, nothing has changed.

The barman is still there, licking his teeth. He has, I unwillingly notice, one of those overdetermined "Italian Stallion" necklaces with the unicorn horn or whatever the hell it's supposed to be dangling amid the fuzzy tuft of his chest hair dickie.

The other patrons are still there, drinking from those narrow-waisted glasses for tap beer you never see anymore. They were all reading the same sports pages or whatever it is you read in a bar for hours on end.

And Peter Jr., presumably, is still there (we have yet to be formally introduced) behind his own paper.

"Hi," I say to the barman. "I was . . . I am the one who was here earlier . . . they . . . you . . . he told me to come back at five for an application."

The barman smiles by forcing his cheeks up without letting his lips or eyes do anything meaningful. Which is to say, he grimaces at me while he dries a glass. After three or four more revolutions of the red rag on the glass, he nods.

"Peter Jr.," he says with more than a little deference toward the still spread *Milwaukee Journal*.

"Johnny Volpe," the man behind the paper says.

"Peter Jr.," Johnny says (man, do these guys like saying their names!), "kid from earlier here to see you."

"Oh I—," I say, involuntarily backpedaling a step or two until I nearly take out one of the waitresses as she drops off a tray of glasses. "I didn't want to bother anybody. I just wanted to grab an application."

"An application!" comes the now bemused voice from behind the paper like the great and powerful Oz. He rustles the *Journal* so it creases desirably and then folds it dramatically once, then twice, and then it's down, and here, finally, is Peter Jr. His demeanor is a cross between Jon Lovitz playing the devil on *SNL* and Marlon Brando in *The Godfather*. "You'd like to fill out an application! How nice. Isn't that nice, Johnny?"

"That is very nice, Peter Jr.," Johnny says. "How very nice."

The waitress I've almost leveled by my crab walking backward puts a hand on the small of my back and gives me a shove toward the bar. "Don't let these goombahs give you no shit," she says. She's got the sweet but rough rasp of a professional smoker. "Go on, sweetheart.

And yous two don't give the kid no more grief. He just wants a job. Right, kid? You here about the job?"

"Yes," I say. "Yes, ma'am. A job."

I like her immensely. She's tough but sweet and reminds me of my godmother, Barb, who once, when babysitting me for the day, took me to a garage sale where she found a powder-blue three-piece suit replete with white piping and stitching. She then dressed me up in it and had my portrait taken at JCPenney, the chosen backdrop a serene pond scene, making me look right at home amid the fuzzy little ducklings and the weeping willow tree. It's a very specific memory but her voice takes me right there.

"All right already," she says. "Saddle up there and talk to Peter Jr."

I take in her order but check it with Johnny, who shrugs deferentially toward Peter Jr., who, like Johnny before, grimaces in a way that seems to mean either he's got arugula stuck between his teeth or he doesn't mind if I should come have a seat. I figure I can't do anything about the arugula, so I tiptoe over and climb up onto one of the big, vinyl, swiveling seats.

"Something to drink?" Peter asks, cordial now that we're on the same side of the bar.

"No, no," I say, "I mean, no, but thank you though."

"Yeah, all right," he says. "So why you want a job here? You go to Marquette?"

"Yes, sir," I say. I've never in my life called anybody sir or ma'am, but I feel like I've unknowingly stepped back into another century where greater decorum is called for.

"That you with your mother from before?" he says. He gestures to the dining room with a big tumbler of something golden brown and icy.

"Yes, sir."

"You have supper with your mother often?"

"You mean here?" I say, glancing around at the bar and back into the empty restaurant. The bubbles in the Wurlitzer float their way up like the dizzying answers I'm trying out in my head. "Um, actually no. This was our second time, but it was great. I had the steak casino and—"

"No," he says, slamming his drink down with more vigor than I'd yet see him do anything. "Your mother, you have supper with her often?

Like since you went off to become a lawyer or a doctor or whatever you're going to school for."

"Oh," I say, "yeah—I mean yes. Yes, sir. I shoot for once a week."

"Once a week, Johnny!" Peter Jr. says. "Once a week!"

"A regular momma's boy!" Johnny says, snapping his red rag in the air like a firecracker.

"Cut the shit, yous two," the waitress says. "Just tell him when to start."

"All right, all right, Cindy," Peter Jr. says, pretend throwing a bar napkin at her in disgust as she turns and leaves. After she goes, he does not say "broads!" but he may as well have. "So. What about it?"

"Sir?" I say.

"Don't 'sir' me again, kid. My name is Peter Jr.," and he shoots back the not insignificant remainder of his drink. "So what about it? When can you start?"

"But," I say, almost physically swallowing the word "sir," "don't you need me to fill out a job application?"

"You just did, kid," Peter Jr. says.

The next Sunday is my first day, but before I begin, Peter Jr. himself gives me the official tour and explanation of duties.

"So," Peter Jr. starts. "That's the kitchen. Those are the cooks.

"That's James Brown. And that's James Brown too. Don't worry about anybody else's names and don't get too attached. There's a lot of turnover in here. Except among the James Browns. One of them is an ex-con. The other is not. Try not to fuck with either of them.

"This is the minestrone. That's the chili. You can have as much chili as you want. The minestrone, not so much. You eat the minestrone, you might be let go.

"This is the reach-in. It's not for you. Don't touch it.

"This is the walk-in. Only that one shelf there with the limes and the lemons and the carrots and the lettuce is for you. For the coleslaw you will make. Don't touch anything else and don't take any more lemons or limes than you need. As for carrots and lettuce, live it up.

"This is dry storage. Bread gets delivered here from Sciortino's. Eat as much as you want, but go easy on the butter.

"What's that? That's a good question. It is called dry storage but there is standing water on the floor. It is still called dry storage. What a good question.

"How do we pass inspection with standing water in dry storage? Boy, oh boy, are you full of good questions. How do I put this so that all of your very, very good questions will be put to bed—I mean rest. Put to rest. Is 'bed' better?"

Here Peter Jr. pauses from his tour and plants his feet firmly on the quite wet floor of dry storage. He looks in the middle distance of the room where a bare bulb illuminates the peeling paint on the cinderblock walls and contemplates his answer. I'm afraid that I am about to star in a reenactment of the jumper cable torture scene from *Rambo* and instinctively cover my testicles.

The room itself is frankly not what one would hope to see/smell/touch/know about before or after eating at such a restaurant as this. There's an open, low, square door that appears to lead to the HVAC system and/or the plumbing where more standing water is evident. All around there are high-water marks on the wall from some worse days. There's a narrow, rusty, metal shelf that I wouldn't put my shoes on, never mind the dozen or so open-ended paper bags of fresh bread. There are upside-down milk crates on the floor—water coming up just over their handles—and upon them are stacked wire shelving units of mostly stuff that, as the name implies, needs to stay dry. Forty-pound bags of sugar and flour. Boxes of penne, boxes of lasagna, boxes of spaghetti. Bags of beans. Bags of rice.

"I would say," Peter Jr. says, finally turning his flat, red-eyed gaze on me, the smell of his Johnny Walker cologne sour in the enclosed space. "I would say, as my father, Peter Sr., would say, that the answer to your very, very, very good question comes down to family. That's right. It's all about family."

He does not, for the record, follow up with any commentary about what might happen to a further questioner's kneecaps should he have more very good questions, but even as dense as I am, I get it.

I have no idea to what extent, but I'm pretty sure that I have just become employed by the mob.

* * *

Donning my black slacks, black shoes, black socks, and a white oxford shirt for my first shift, I feel like the skinny, toast-eating Elwood Blues of the Blues Brothers, but instead of my surly and vulgar ex-con brother Jake at my side, I have a small fleet of almost-elderly-but-still-tough-and-young-enough women. Turns out, other than the cooks and the bar staff, I am the lone male employee in the restaurant. I am the busboy.

Initially, I think being the only college kid as well as the only guy in the restaurant is going to be a problem, and as I try to settle into my first shift of coleslaw duty, taking a raft of whole carrots and shoving them grotesquely into the spinning disk shredder of a food processor, I contemplate letting my stack of carrots just drop to the floor and bolting for the nearest exit, but, for some reason, I don't.

Part of me thinks they know where to find me. But another part of me is just interested in the place, the people—and the food, too. It's simple and unadorned, but super homey. Food that is inspired, I learn soon, by Peter Jr.'s grandmother, Katherine. Miss Katie.

Of course.

No wonder Peter Jr. was grilling me about suppers and how often I had them with my mother. He named his own restaurant after his grandma.

The duties of the job are varied and shift pretty quickly, which suits my fleeting attention span. One minute I'm shredding carrots and thinking about my prematurely lost masculinity, the next I'm rinsing off heads of lettuce and spearing them with a tiny but frighteningly sharp paring knife, slicing out their cores. After that, service begins, and my job shifts from the back of house to the front, and the change is as palpable as going from the rough and grimy wings of a theater to the proper stage itself.

Only I'm not an actor in this play. I'm just there to keep the waters and coffees full and the dirty dishes clear.

But since this is a Sunday when the Milwaukee Bucks or Brewers aren't playing, we're dead. And so, because this is Wisconsin, where, back then, if you didn't earn enough tips in a restaurant job you were paid minimum wage—the difference between making, believe it or not, $2.13 an hour (the state's going rate for tip-earning staff) or nearly seven bucks an hour—and so, thanks to the sweet all-lady crew of Miss Katie's dining room, I end up spending the last two years of college mostly not working at the restaurant, where I nonetheless got paid to eat chili (but not minestrone) and amazing Italian bread with just a little butter and to do my homework in the booth back by the kitchen.

The ladies would hang out with me as I read my Chaucer or scanned lines of Shakespeare, clicking their tongues, shaking their heads, always smoking their Winstons or Virginia Slims, saying, "Good for you, kid, good for you. How far you'll go!"

And though I would leave soon enough for Boston and graduate school, in many ways I'd never work at a sweeter or more supportive job. Not because it was almost always not working—on game days we'd be crushed with baseball or basketball fans from wall to wall for hours, both before and after games. And not because they were all kind and grandmotherly—neither Peter Jr. nor Johnny Volpe ever warmed up to me beyond simply letting me work in the dining room, in which neither of them ever intentionally set foot. And not because I was particularly good at any part of my job. Rather, it was because when there was nothing to do, the ladies let me do the something I had to, and when there was lots to do, there was nothing easier than busting my ass as their busboy to pay back the hours that they let me while away like a young goatherd from another century.

Once I rented an apartment in Boston and figured out how thin my financial aid was going to be spread—mediocre students like me were baited with the prospect of earning teaching fellowships not in our first semesters but maybe later—I wasn't even a week into grad school when I stumbled down the block in my Fenway neighborhood to apply for whatever restaurant gig I could get.

As it turned out, just two blocks from my apartment was a "family" restaurant named Thornton's Fenway Grill. It was also run by a tag-team of guys, brothers this time, Bud and Marty, a rough-hewn pair of Michigander transplants. They vetted everybody who came and went in their namesake greasy spoon. Since I was willing to work pretty much whenever I didn't have class, the only question they really wanted to know was if I willing to work for tips.

"Of course!" I exclaim like the Wisconsin hayseed I was.

Bud and Marty stroke their weak ponytails and scraggly beards. Bud crosses and recrosses his legs, in part, I'm sure, to show off the shiny silver tips that adorn his cowboy boots, pointy enough to kill a cockroach caught in a soup can.

"We mean," Bud says, leaning forward as though in confidence despite the fact that the whole rest of the lunch staff is sitting at the table behind us drinking beer after their shift. He coughs and leans in further. "We mean, are you willing to work only for tips?"

"Oh," I say.

I had just come from an honest job working for the mob and here I am interviewing to work illegally for a couple of guys from Grosse Pointe.

"Did you say anything about the free beer?" Marty chimes in.

"The free beer?" Bud says.

"The free fucking beer, Bud."

"Jesus Christ, Marty-the-one-man-party," Bud says. "What do we know about this kid? Maybe he don't even drink beer."

"Who the fuck wants to work in a restaurant what don't drink beer?"

"Good one, Marty," comes a comment from the staff at the table behind us. The fact that the *r* in his name doesn't even make a passing appearance in the way he says it leads me to believe that, regardless of me and the owners, I won't be surrounded by Midwesterners here.

"You drink beer, kid?" Bud says.

"Sure," I say. "I guess."

"Fancy beer or regular beer?" Even though Bud's from Michigan, he's clearly been here long enough so that regular comes out reg-you-lah.

Immediately I tense up. Imitating accents has always been my own strange variety of Tourette's. Somebody with a prominent accent says something and I can't help but parrot it back. I roll my lips and bite my mouth shut. I know "fancy" is the wrong answer and I sure as hell can't say "regyoulah" but I really, really want to.

I try a shrug. I tug my ear. I squirm this way and that in my seat. I have to answer.

"I'm not paticulah," I say, but then cough to cover it up.

Both Bud and Marty squint, on to me, but then let it pass. They look around at the rest of the staff behind them, out the windows at the Russian ladies with their babushka hats walking by on Peterborough Street, at the Bob Seeger and the Silver Bullet Band poster aglow with its Christmas light corona around the frame.

"Reason we need to fucking know," Bud says, gesturing wildly as though I have just accused him of sedition or sodomy or some other dark crime.

"Yeah," Marty says. "Reason we need to know is—"

"Reason we need to know," Bud says, hitting Marty on the shoulder, "is because of that fuckwad whatshisname—"

"Yeah, whatshisname."

"The guy what drank all our good beer."

"The expensive shit. The Samuel Smith Nutbrown."

"That's the stuff."

"Fucking expensive goddamned beer."

"Sure as shit is."

"Used to be," Marty says, glancing at his reflection in the picture window, clearly trying to see what he thinks of himself in his weak goatee. "Used to be, this place was like an extension of my living room."

"Our living room," Bud corrects.

"Our living room," Marty assents. "Shift's over, everybody hangs around, parties with us and has as many beers as they want."

"There was the idea, see," Bud says, squeezing out his ponytail as though it were wet instead of just greasy.

"That the place was for family," Marty says. "The family that works here. And then this guy—"

"Whatshisfuckingname," Bud says.

"Toby or some shit," Marty says "Whatevah."

"Yeah," Bud says. "Toby or whatevah."

I am biting my lips so hard at this point that the coppery rich taste of blood is on my tongue.

"Anyway," Marty says, "so long as you don't think you're going to get to drink fucking Samuel Smith Nutbrown after work—"

"Or Harpoon," Bud says.

"Or Harpoon," Marty says.

"Or get a paycheck," Bud slips in.

"Then we guess you're hired." Marty shakes his hair out as though there's a Fabio-esque train of locks behind him.

"Yeah," Bud says, "you're hired. I mean not W-2 hired. But, you know, you can come here and do stuff for tips and shit. What're you drinking?"

"Oh God," I say and time comes to a standstill. Do I say water and end up with the waistband of my underwear over my head? Do I say a Coke and that I've got to study tonight (and end up the same way as in scenario 1 but possibly also floating face down in the Charles)? Do I say a light beer even though I'd rather drink liquid wrung from a bar towel? I know I sure as hell can't say anything about a regulah beer, and I think strategically for a second and turn to the staff table to see what they're drinking but it's all just beer in glasses—no tell-tale bottles or cans in sight. I decide to risk it all.

"I'll have what you're having!" I say, having made sure that there was no nutbrown anything in the glasses Bud and Marty were drinking out of.

Bud and Marty go a little more silent than they already just were. A line has either been toed or crossed. They both do this thing with their mouths that makes me wonder whether they're trying to tie the stems of cherries with their tongues, but finally Marty speaks.

"I guess this one time."

Bud shrugs. "Yeah, what the fuck. This one time," he says, and he gestures with three fingers to Rock, the bartender who has been as peripatetic behind the bar as a radio dial being turned by somebody trying but not finding anything he wants to listen to. Rock raises his eyebrows to make sure, but Bud and Marty nod, and a minute later, Rock is back.

"Here you go, boys," Rock says, dropping off three highball glasses. "Three Malibu and Diet Cokes with Grenadine and extra cherries."

The three of us clink glasses, and just before I take a sip, Bud speaks up. "But don't get used to this. This is the drink of the ruling class. The management."

"The landed genrtries," Marty says, taking a loud, sloppy sip through his cocktail straw.

All I say is Cheers and I take a drink and immediately have to vault my mouth shut to not spit the drink out. It tastes like a kiddie cocktail doused with crushed aspirin and suntan lotion served in a partially de-flated beach ball. I wonder how badly I would hurt their feelings if I give this back and ask for a good old-fashioned Buck Rogers instead.

Mercifully, Bud and Marty wander off to mingle with the various womenfolk at the other end of the bar and I'm left alone.

Behind me there's the sound of a couple of chairs scraping the tile floor so that the rest of the staff can now take in their new quarry.

A severely unimpressed woman with a bun as high and tight as the pommel of an English saddle grimaces in my general direction.

"I'm Lydia," she says, but it comes out as Lid-ee-er, and I quickly shift my gaze to the next person.

A small, spritely girl with glitter for eye shadow and a feathered ear-ring winks at me. She takes a dramatic breath and then says her name dramatically on the exhale. "Mandy!"

A Billy Idol-esque guy across the table from her nods in a backward fashion and scrunches his lips at me. "Malibu and Diet," he says. "You a fucking quee-ah?"

"David," Mandy scolds. "What the fuck is the matter with you?"

"What?!" David says. "He's drinking fucking Malibu. Only so many paths that can lead down."

I can't even possibly be upset—no matter my sexual orientation—since David is clearly from around here and the silly music of the way he says quee-ah only makes me happy.

Just then the guy next to him bristles as though to shake off the dust and pine needles that have been quietly accumulating on his floor-length leather trench coat over the many months he may have been sit-ting here.

"I'm Fu Man," he says, and his eyes flare and then go back to dis-trustful slits. "I have rocks in my pockets."

I nod as neutrally as possible. The name must be more aspirational than descriptive as his facial hair is, at best, a scraggly rumor of a mous-tache. Still, he seems unstable and certainly in need of no lip from a guy such as me.

"That's cool," I say. "Rocks. Never know. Ballast . . ."

He nods philosophically, and I think we will be friends.

"What's up?" the guy next to him says. He's apologetically tall—you can tell even by the way he sits with his shoulders slouched to take up less space—but he's got bright, warm eyes. "I'm Greg. And no, I don't want to talk about Poughkeepsie."

"What about Poughkeepsie, Greg?" David says.

"Jesus fucking Christ, Dave!" Greg says, slamming a tiny bit of his beer as though it'll quench whatever thirst ails him. "No, I'm not Button Briggs or Tommy fucking Boggs or goddamned Ed Wood even though I was born in the same damned hospital. And no, I didn't get in to Vassar. I didn't even apply to Vassar, Dave. Well, I did, but I didn't put enough postage on the envelope. It was a really oddly sized envelope. Enough about Poughkeepsie!"

"Oh, Greg," David says, finally dropping the attitude. "You're so beautiful. Anyway," he says, his hard eyes back on me. "I'm Dave. Dave DeCook. Welcome to the occupation."

"Thanks, man," I say. And just like that I feel like I've found a new family.

It will be two years, by the way, before I learn that "DeCook" is not Dave's last name but rather, simply, his job. He's de cook.

Chapter Ten

PART-TIME PIRATES

WHEN JENAE AND I LAND IN COLUMBUS, OHIO, SO I CAN DO yet more graduate work, this time I do manage to get a teaching fellowship, but it only pays ten grand per year, so I get a job at a knockoff Wolfgang Puck restaurant. The owners were even German and so couldn't possibly have been unaware of Chef Puck, arguably the first international celebrity chef. They went so far as to copy the font and logo of his seminal restaurant, Spago. In order, I guess, to minimize effort and to maximize accidental crossover appeal, all they did was add an *i* to the name; hence, from Beverly Hills's Spago, Columbus's Spagio was born.

In almost every way, it was the worst restaurant I had ever worked in and, I hope, ever will. First off, its concept was super muddled and therefore without a real identity. Despite the fact that the owner/chef, Hubert, and his wife, Helga, in charge of front-of-house operations, and also his sous chef, Gösta, were all German nationals, the supposed concept of the restaurant was "European/Pacific Rim Cuisine." Back then, I wasn't sure that was a thing. Now I know it's not a thing.

To this day, remarkably, Spagio is still in operation, and more remarkably, they still identify/advertise themselves as a European/Pacific Rim restaurant, despite the fact that the Weiner schnitzel and the spätzle are never going to be confused with *anything* from East Asia (though purportedly they still have that one jar of kimchi way back in the reach-in just in case anybody wants some dank, pickled vegetables other than sauerkraut).

Still, Spagio offered both the best and the worst of potential experiences in the restaurant universe.

On the upside, as in any restaurant, I got to work closely with people from all over the world. And I mostly don't mean the stereotype-enforcing, humorless, exacting, irascible Germans. I mean immigrants from South and Central America who were, without exception, hard workers, generous friends, and eager and grateful new citizens.

My favorite was easily Jaime. He and I were about the same age, working more or less the same shifts in the same restaurant, but here he was, about a million miles away from his native Nicaragua. He worked the wood-fired brick oven (because what better fuses German and Pacific Rim cuisines than pizza?), and every now and then he and I would mutually draw the short straws and have to close our respective sections by ourselves.

Spagio, by the way, was the kind of restaurant that would be open nearly all hours of the day or night despite the fact that the beginnings and ends of days were nearly always dead. But their logic was, we can shut down the proper kitchen (with the German chefs) but leave open the pizza oven (with the undocumented Nicaraguan cook making pennies on the dollar). And in the dining room, we can send all of our lead servers home, as they've already made the most money in the shortest amount of time, but let our newbies grind it out to wait on whatever dregs of the night might wander into a schizophrenic German restaurant advertising Asian food that's open until two in the morning in the third most relevant insurance city in Ohio.

Even with all of my ludicrous closing duties (clearing and resetting all tables, refilling salt and pepper shakers, rolling at least fifty silverware setups, vacuuming the acre of carpet in the dining room, mopping and restocking the server stations, and cleaning both bathrooms and also the bar and even behind the bar)—even with all of that—Jaime and I had a fair amount of time to hang out and work on our mutual interests in learning each other's language.

Unfortunately for both of us, neither of us was all that good in foreign language instruction. I did work for a while in Boston at Berlitz (the McDonald's of language instruction) but all it left me with was the ability to encourage parroting to the tune of "Is this a pencil? Yes, this is a pencil!" Somehow, we managed to learn a good deal about each other that I think we both found fascinating and/or hard to believe.

For me, it was all he had been through just to get to the United States and the insane amount of work he was constantly doing in order to help take care of his family back home. For him, I think it was that someone would go to high school, then college, and then graduate school, and then more graduate school, all just ostensibly to become a better writer and teacher of writing.

Jaime would stare at me, bleary-eyed from the heat of the brick oven and his usual sixteen-hour workdays. "Write-ing?" he would say, imitating my scribbling on an order pad. "This what you go school for many years?"

I was afraid he thought that writing meant handwriting or calligraphy, and I was even more afraid to try to elaborate toward something arrogant and explain that it was literary writing I was after. What, after all, would I have said? *No, no. Not writing like with your hand. Writing like Gabriel García Márquez! Writing like Borges!*

Instead, of course, I just nodded and shrugged. "Yes," I said. "I want to write better."

But, as odd as it may seem to folks who haven't worked in the industry, I also wanted to become better at working in a restaurant. Spagio didn't have many servers I considered as mentors, but there were a couple of notable exceptions.

The first was a waiter from Portugal named Paulo. Paulo was a short, wiry, exuberant man who described himself as a *"profesionál!"* He insisted that he was not just some schlub of a waiter like me and pretty much everybody else who worked at Spagio, but this was his career, his *"pasión!"* (Whether it was in Portuguese or Spanish or English, pretty much everything Paulo said needed to be in italics and with an accent on the last syllable and an exclamation point to get anywhere close to the kind of stress and emphasis he would put on whatever he said.) In most ways, Paulo taught me that there is a way to regard the powers that be in any given restaurant. Go to any mid- to high-end restaurant anywhere in the world, and I promise you'll find more than a few of his old school there. Basically, it is a mindset that regards the hierarchy of the restaurant above all else, and it does so with continental—or, more specifically, French—erudition.

I don't care where any given executive chef is from, when (not if) he is dressing down his cooks or his dishwashers or his front-of-house staff, there will always be a contingent among them who reply, no matter what the question, "Oui, Chef!" in the same manner a new recruit at boot camp is going to shout back, "Sir! Yes, sir!" no matter how casual the question.

But what else Paulo taught me was that that level of formality and deference wasn't actually just an acknowledgment of servitude. It was also a condemnation of the chef's need for reassurance in the face of grave self-doubt. In the case of Hubert or Gösta, it typically went something like this: "Ven I say ve have no more Weiner schnitzel and zen you go and you order more Weiner schitzel I tink I am losing my mind because did I not already say unto you dat ve have no more Weiner schnitzel so do I actually need to say unto you dat dat means you cannot order no more Weiner schnitzel?"

The other servers stare, baffled, indifferent, bored at Gösta/Hubert, but Paulo—Paulo lets just about one beat too many pass, and then, with I swear to God a click of his heels, he shouts, "Oui, Chef!" and order is restored.

"Oui, Chef!" doesn't mean *I'm sorry* or *You're right* or *That was a good talk* or *Perhaps we can nuance this so all sides are feeling validated*. No. When the subtext of what the offending chef or cook is saying, *I know you don't want to say or do anything more that would cause me to kill myself and/or others more than I already do, but not saying anything isn't good enough and I've already talked/screamed more than can be reasonably replied to, so just stick with the script and we can all get back to work*, "Oui, Chef!" is all there is to say.

In chef memoirs or on Food Network shows, working in a kitchen or a restaurant more broadly is often compared to being a part of a brigade or the crew of a naval vessel. Though it might seem like a stretch to compare making calamari in a big-city café with being attacked by a giant squid while sailing the high seas above several miles of rough water on a wooden trireme, both do actually share that insane devotion to the patriarchy and the captain himself.

I'm glad to say that, for the most part, this is a back-of-house thing. In the front of house, among servers and whatnot—perhaps because we get out a bit more, work shorter shifts, or have to talk with guests who are, usually, human beings and not peg-legged, swarthy buccaneers—things tend to be a bit more relaxed. Not true, however, for Paulo, which kind of made him my hero.

The other thing that Paulo taught me is that, except in the above scenario, whenever things are going your way, there is no reason to be meek about it. You get out of a close call with management? Get a hefty tip? Convince somebody to order a spendy bottle of wine? Get a cheap-seeming couple to leave the restaurant because we don't serve Coors? Win a bout of rock-paper-scissors for who has to clean out the sanitary napkin boxes in the ladies room?

According to Paulo, there is only one way to reply. You say/shout, in that Portuguese way only he could really pull off, with every syllable stressed equally and emphatically, "THAT IS WHAT I AM TALKING ABOUT!"

In the same universe at Spagio but in a different orbit, there was Phillip. There wasn't anything globally interesting or exotic about Phillip. They're so different from each other, I don't think Paulo and Phillip could reasonably appear in the same jury or even the same urgent care waiting room. Where Paulo was full of *pasión*, Phillip was sedate and mild mannered. The most interesting thing about Phillip was that he was from Wapakoneta, Ohio, and that once, at the Ohio State Fair, he had stood before the likeness of Neil Armstrong, his hometown hero, rendered entirely in butter.

But Phillip, like Paulo, was a *profesionál.*

If it rings any bells, by the way, I think of the word *profesionál* in the same way I remember reading about how Hemingway talked about bullfighting *aficionados.* In other words, we aren't talking about the standard-issue bunch of louts just fucking around. There's no such thing as part-time bullfighters. This worldview does not admit semi-professional pirates. We're talking about people who are fully committed to whatever they do. People who are here not because, in the case of the restaurant business, they think it's a nice sideline of work while they pursue acting or pottery-throwing or Irish step-dancing fame, but rather they're here because they believe restaurants to be the ultimate venue for art, the theater of life where all aspects of beauty, hospitality, industry, cookery, and commerce intersect. And, when they're at their best, they do so without a single note of effort. But that takes real *profesionales.*

Phillip, despite the fact that he was, at least for the time being, stuck at Spagio, was among the best waiters/hosts/sommeliers I'd ever seen. He wasn't, however, always at Spagio. Sometimes he was at Spagio's private restaurant next door, Aubergine (French, I think, for eggplant).

It was a lovely name—not ripping off anybody that I know of—and so was the restaurant. It was right next to the Spagio mothership in Grandview Heights, but it had its own entrance though it had no name on its door.

If you'd have asked me before I set foot in the place if I thought it was a good idea, I'd have said, sight unseen, decidedly no. Spagio was rarely all that busy, and never enough so that I would've thought they needed more or another room. If you'd have asked me if I thought it was a stupid, pretentious idea, I'd have said decidedly yes.

But I was wrong.

Whereas Spagio was a tacky, vague, and bland imitation restaurant desperately trying to masquerade as a fine dining establishment, Aubergine didn't try to do anything other than what its guests wanted it to be. Which is to say, if you were prepared to book a space, a waiter, a sommelier, and a chef, you were likely not doing it so that your guests could eat chicken nuggets in the shapes of their favorite dinosaurs while they chugged Mr. Pibb until it shot ceremoniously out of their noses. No, if you were going to book a private event at a private restaurant, you were damned well going to pay for it, and there's no way you didn't know that in advance.

I imagine this is quite often a perfect recipe for an unrepentantly pretentious evening among the world's preeminent butt sniffers, but I never got jaded because I only got to work there a couple of times, and both of them were transformative.

The first time, I showed up to the back entrance of the kitchen at four, and unlike the generic industrial kitchen of Spagio, the private kitchen of Aubergine resembled rooms I'd only ever seen on *Lifestyles of the Rich and Famous with Robin Leach* or the more recent and imminently more hateable *House Hunters International*. Instead of gritty stainless steel everywhere, there were maple cabinets with stained glass doors, stacks of bone white china plates and bowls inside, quiet and pious as ranks of nuns. Instead of the black-bottomed, battered aluminum sheet pans and pots, the only cooking vessels I saw were hammered warm-gold copper. And instead of the mishmash of the piratical crew of cooks, Chef Hubert himself was manning the stove, happy as only a chef can be when he knows the night ahead of him has already been paid in full.

In the dining room, I found Phillip wearing the typical waiter's black pants and shoes, but then, from the waist up, just a white cotton T-shirt—his starched and pressed dress shirt was on a hanger on a sconce—and he was hard if not slow at work, cleaning and polishing wine glasses. There were enough glasses on the table to serve all of Spagio on one of its key party/live jazz nights, and yet I knew that the group we were preparing for was only a group of twelve.

And while the majority of the night was a complete blur with me only ever doing whatever Phillip told me to do, I remember the beginning, and I also remember the end.

At the beginning of the shift, the attention to the detail of the table setting dumbfounded me.

First off, the silverware. Not good enough to be clean, we polished each and every stick of flatware and inspected them all not for dirt or leftover food, which was all I'd ever done theretofore, but here even the slightest ghost of a water spot was unacceptable.

The glassware was even more intense. Phillip would drag an already clean rack of wine glasses from the dishwasher and, one by one, he'd take each from the rack and hold it over a pot of hot water until the inside of the bell of the glass was steamed over, and only then would he take his microfiber cloth and meticulously polish and shine the inside and then the outside and then the stem and then, finally, even the base of each glass.

If I ever kill somebody, Phillip is my go-to man for taking care of the crime scene.

The end of the night was a different kind of shock. Different because it was at once utterly simple yet completely revelatory. The bulk of the evening had been a pageant of Hubert introducing the various dishes he had prepared—Chilean sea bass this, roast duck that, suckling pig whatnot—and Phillip describing the wine he had selected to accompany the food—this cabernet franc with notes of blackberry and Tellicherry pepper, that pinot noir with its jammy and mysterious overtones, this rustic whathaveyou to boost the earthiness of the rutabaga—and so on. But the end was—you'll have to forgive me—all about paring back.

In little glasses shaped like crystal tulips, Phillip poured a liquid that looked like it was somehow both perfectly clear and also golden.

"Poire Williams," he said, simply and without elaboration.

"Und your pear," Hubert said, as unpoetic as though Schwarzenegger was speaking through him, "poached with ginger and peppercorns, suspended in white wine aspic."

Compared to the basilica of a kitchen it came from, the dessert looked pretty underwhelming. It just looked like half of a peeled pear suspended in Jell-O.

Phillip insisted on me getting to try at least this last pairing, and though I didn't really want to, when I did, the combination was like an absolute explosion of all things pear. Theretofore, my experience of pears only involved the grade school cracking of the seal of a tin of pears right before somebody would invariably break a box of pencils over my head.

This was light, aromatic, delicate, intense, and—as best I can describe it—zoomy. The flavors just kept coming at me. Ginger one moment, then the hot rush of the brandy (Poire Williams is a pear brandy), then the spicy peppercorns, then the soothing sweetness of the pear itself, then the cooling ease of the aspic, which, yeah, is basically flavorless Jell-O.

It was haute cuisine, but it was also still the nineties.

From Spagio I gladly took my leave to take a crack at what became my first foray into working for a restaurant group. Not quite a chain, and definitely not just working for mom and pop, working for a restaurant group is kind of like landing a spot on an up-and-coming farm team. It ain't the majors, to be sure, but from here it can lead to bigger and better things. More important, restaurant groups tend to have dialed themselves in to the pulse and demographic of their given community and, for the time being anyway, enjoy the kind of popularity that seems both ludicrous and somehow maintainable.

The group in question was the Cameron Mitchell chain of restaurants. Now there are over fifty-seven Mitchell group–owned restaurants in more than twelve states. Then there were just three: Mitchell's Steakhouse, the Cap City Diner, and the Columbus Fish Market. The latter was the newest and, more important, the only one hiring.

Before I began my training at the Fish Market—despite its name it was actually an upscale, fine-ish dining restaurant featuring the freshest fish flown in, supposedly, daily from all over the world—before I could start, I had to read a book. It was called *Raving Fans*, and though it was barely over a hundred pages long and had dubiously wide margins, I found myself inexplicably smitten. At the center of the book—and the philosophy that the Mitchell group wanted all of its employees to espouse—was the notion that, and I quote, "The answer is Yes! What is the question?" (Mitchell loved that phrase so much that he just wrote his own memoir and gave it that as its title. No higher praise than—what was it? Plagiarism?) At any rate, the idea behind the book was simple: you try to out-service the competition.

Everybody in business, no matter what the trade, is basically out to move more widgets than the next guy. And, anymore, everybody's widgets are pretty much all made or processed or imported by the same megacompany, so what's a business to do to get a leg up?

You can only cut prices by so much, and you can only source product that's so cheap, before both tactics reflect back on your business and, if you're in the restaurant business, you become known as the Wal-Mart of dining.

So, the CFM decided to try to go the other way.

Best product. Best service.

They were willing to bet that in a decent economy, people would, within reason, be willing to pay more to get more.

And since this was a fish place, it had to start there. It wasn't flashy or exotic, but it was lovely and always fresh. (Pro tip: if fish smells "fishy" it's not because it's fish but because it's bad or old—except salmon. Salmon always tastes and smells like bad fish.) The restaurant had a glass-faced counter at the front of the dining room where, if you wanted, you could literally pick out the fish you wanted and tell the chef how to prepare it. There was as much transparency as is possible in terms of what was being sold and why the price was what it was. Fish, unlike, say, cows or pigs, aren't always in equal supply. Hell, you can source beef or pork from more than one country and it'll keep pretty much indefinitely in freezers all over the world and purveyors can scour global markets to find the best product at the best price.

With fish, well, you're talking more often than not about the fishy equivalent of a beef steak or a pork loin—and, most experts agree, it can't be frozen or it'll lose both texture and flavor, and so it's got to get from water to table as quickly as possible. Oddly enough, land-locked Columbus, Ohio, could draw on purveyors of fresh fish from most oceans and all over the country. But still, when you see "market price" on a menu for whatever the fish special is, that's because they, too, have to pay whatever the market price is and add a percent or two as well to keep the lights on and baby in new shoes.

The food was great and the restaurant was almost always busy, and though we were more cogs in a corporate machine than in any restaurant I'd worked in yet, we were oddly empowered by that "the answer is yes" mindset to do whatever it took to make sure our guests were not only satisfied but thrilled with their experience.

Not everybody at the CFM went this far, but for me, with my roots in my mom's over-the-top customer service in her flower shop, finally I had found a place where I could be really enthusiastic and creative and could just kind of go nuts to make sure everybody was having an amazing time.

I served one guest who was bummed that we didn't have Barq's Root Beer, so I ran—literally—to the gas station a quarter of a mile away to get him a can.

Another guest at a large party of enthusiastic fish eaters joked that she just wanted something from the Burger King that they passed on the way to the restaurant, so I jogged over and picked her up a couple of Whoppers.

With every check, we included a guest experience survey, and if somebody included their name and address, I'd always handwrite a thank-you note to them.

Whereas at Spagio we would seek out ways to exclude diners—no cheap domestic beers, no ketchup, no microwave, no kids' food—at the Fish Market we were able to strive to make their time with us not just fancy but truly hospitable. Which is to say, we knew our job was to make sure everybody had a great time no matter who they were or why they joined us.

And beyond that, it was *fun*. We were out to make sure our guests were having a great time, and so as long as everything was aimed in that ultimate direction, we could entertain ourselves too.

We had an amateur opera singer who would stand on a chair at least once a night to belt out "Happy Birthday" in Italian to anybody we thought could stand it.

We were discouraged from smoking but were encouraged to take nonsmoking breaks on the loading deck out back whenever it was opportune. One guy, another Dave, even brought his own folding camp chair and a milk crate for his footrest that he would set up back there and lounge for forty-five-second "power breaks" in between seatings.

Dave was also the guy who really instilled in me the value of, whenever you could get away with it, just screwing around.

When we'd be training someone new, he'd make sure to send them to the basement aquarium to help harvest fish for the dinner shift (we had neither a basement nor an aquarium, of course), or when things were about to get really hectic before a weekend shift, he'd take a bunch of disposable cups and cordon off a section of the floor right in the middle of the dining room as though there had been a chemical spill or a crime scene and wait to see how many people obeyed the command of those little cups. Nobody but our manager would ever cross the line.

Really, it was a great place to work, and I was happy to stick it out there for three years as a part-timer while I finished grad school and began teaching. But in the end, I have to admit, it got a little old slinging the same varieties of fish the same three or four ways.

After Ohio, we moved to Salt Lake City for, yes, yet more graduate study, and accordingly, more restaurant work too. My experience with the regional restaurant group model was so good and the money was so dependable, however, that my main tactic in Utah was to find the Cameron Mitchell equivalent.

That would be the Gastronomy group, and they ran a handful of popular restaurants around a similar concept as the Mitchell folks. Fresh fish. Fresh grilled meats. Great service. I landed a spot at a literal underground, basement restaurant called, of all things in Utah, the New Yorker. It was a solid, stalwart of a place that had turned out consistent, good, steakhouse/supper club food for nearly thirty years by the time I joined them, and though their menu lacked any real enthusiasm other than fresh and almost-top-end product, it was a good place to work. It was also the longest I worked at any one restaurant, which gave me some perspective on the industry my other food service jobs hadn't.

First of all, it was the first restaurant I worked in that had lasted long enough to have been around for the strange and raucous eighties, when, if rumors are to be believed, it was common for both front- and back-of house employees—and diners—to routinely involve at least a couple of bumps of cocaine in the course of any given evening, and said bumps were likely to lead to more bumps and/or bumping and grinding in the walk-in cooler, in the dry storage, in the men's/women's rooms, in the coat check room, and so forth.

Since a debacle on a fateful New Year's Eve at Spagio (I drank way too much champagne from the glasses I cleared from my tables and woke up around 3:00 a.m. in the large planter out back, an enraged Helga my first charming vision of the New Year), I didn't mess around with drinking at work, never mind snorting lines or popping pills. Folks still did it, to be sure, and I'm sure they still do, but at most restaurants these days, it's less about living any kind of lush life and more about the management of the chronic pain that goes hand in hand with working what is, to your body anyway, a manual labor job. Slipped disks, torn ligaments, tennis elbow, tendinitis, migraines galore—it hurts to work like this. And, as I was learning plenty well, the older you get, the more and longer it hurts.

And because the NY'er had been around as long as it had, there were more so-called lifers there than in any place I'd ever been. There were also more grown-ups than in any restaurant I'd worked thus far. There were single and paired-up parents, real estate agents, insurance agents, flight attendants, and, as was my case, also the usual spate of college students who worked there. But, unlike any other place I'd been, the full-timers' ranks were closed off and protective of one another to a ferocious degree. I had to work months and months of shitty lunch shifts where I'd often wait on no more than one or two tables of two, leaving me with little more than a ten-dollar bill to prove I even went to work, before I was cleared to work the night shifts. And even then, it was a couple of years before I was ever even scheduled for night shifts. Most days—my preferred ones were Wednesdays, Fridays, and Saturdays—I'd just show up right before everybody was about to clock in and see if anybody was too hungover/sore/tired/bored/horny to work that night and I'd swoop their shift.

Even so, if, say, Carole—the server with the most years clocked at the New Yorker—would punt her shift to me because she had to have a bunion checked out, I'd get her tables, but the guests assigned to them would suddenly slide to Ali's or Matt's sections (I was Matthew there because another Matt was there first) and I'd be left with the walk-ins.

That began to change when I finally learned one very important and essential truth: No matter what the employee hierarchy plan says, WHOSOEVER ANSWERS THE PHONES IS IN CHARGE.

This is as true for restaurants as it is for sales as it is for academia. The host or receptionist or administrative assistant or whatever their equivalent is, in whatever line of work you can think of, though they might not get paid like it or respected for it, they're the real bosses running the show, and the quicker you figure that out, the better your life is going to be.

And I don't mean to be pragmatic or careerist about that. The fact of the matter is that very often their jobs are also the ones that receive the most frontline abuse, that get the lowest pay, that get the least recognition or respect and have the highest turnover. But you try running a restaurant or sales floor or college or whatnot without somebody on that sharp end of the business and you'll all be up to your ankles in your own tears within a week.

It was also at the New Yorker that I finally realized there are folks for whom the job of host is a public profile position as much as it is a place to earn a wage.

We didn't have one at Miss Katie's or at Thornton's, and the one we had at the Fish Market was great but so young and obsessed with flirting that it was hard to tell what her job was other than to wear gartered stockings and to repeat your last names and the number of people in your group. At the New Yorker, however, we had the absolute host to end all hosts. Allison.

I have no idea what Allison got paid, but she looked like she was ready to model and/or release her own line on Fashion Week in Milan.

She was tall and willowy and always impeccably dressed, and she could quote Wim Wenders and Chevy Chase movies equally well, and she was also always as ready to dish on whatever dirt was going down as she was ready to give you a hug and say in that strange-for-Utah mid-Atlantic way "Daaarrrlllliiiing!" no matter who you were.

Everybody—gay, straight, young, old, rich, poor, you name it—we all loved Allison and Allison loved us all. It wasn't so much like Allison worked at the New Yorker, though she very much did—I knew who the owners were and she was decidedly not among them. She was just a beautiful single mom who worked paycheck to paycheck like the rest of us—but the way Allison carried herself, it felt like the New Yorker worked for her.

She was, as I have never known before, the maître d'hotel. Which means, naturally, not simply host or greeter or secretary but rather, magisterially, Master of the House.

The more I think about it, the more I realize that restaurants are less about food and more about people. Not just the folks working on any given shift, but the folks who make up the restaurant. I think it was Tom Colicchio of *Top Chef* (née Craft) fame who said that you go to a restaurant the first time because you heard the food was good but you keep going back because of the service. The deeper, real truth, I think, is that you keep going back because you want to be a part of the restaurant. Nobody in their heart of hearts wants to be a regular at Houlihan's or Hooters or some other mall parking lot abomination. We all, down deep, want to be regulars at someplace aspirational—someplace that we secretly know is a little fancier than we actually are because it makes us feel, simply, special.

If all you have for clientele are douchebag, corporate-account as-shats, no matter what else you do, that's going to be who you are as a restaurant and that's how you're going to feel as a diner there. If all you have for cooks and/or servers are people who are waiting for their big break in Hollywood/Nashville/Broadway, then similarly that's all you're going to be. Temps and posers.

At the New Yorker, we had a really lovely mix of people who would never have otherwise gotten together—everyone from the ruling elites of the LDS Church to the self-appointed Basement Remodeling King of SLC to Keith Lockhart, then the director of the Utah Symphony

Orchestra—and they all came together over the course of many, many dinners in large part, I believe, to see Allison.

But we had a special staff there too, with more solid people working there year after year than anywhere else I'd ever worked.

There was a guy named Kim, who managed to make a career out of being a lunch-shift line cook and did so in such a way that he would save enough every year for him and his partner to spend a week in Oahu. Only a week, you might say, but it was always enough for him to come back so addled with the charms of the island that he'd play nothing but ukulele music for the next several weeks.

And then there was Andy, a self-identifying drug addict and former dealer who had moved on to drinking a fifth of hard liquor every night, but that didn't stop him from coming to work on the line every day with an attitude sunshiny enough for him to spend ten to twelve hours slinging steaks while belting out John Denver songs where he'd replace all salient nouns with female genitalia. (But in a good way, if you can imagine that.)

And then there was Bob, the grumpiest, and also my favorite, who was always equally willing to either make you his special not-on-the-menu BBQ chicken sandwich with lettuce and diced pickles or confront you if you were going to do something he knew was morally wrong. If you thought you were too busy to put a dirty dish in the dish pit or had too many tables to run somebody else's food or tried to sneak a smoke break when really you should help Andy and Gabi peel potatoes, Bob was always there to simply but philosophically demand, *"What the fuck is the matter with you?"*

It's a common enough phrase in the world and in restaurants, but unlike everywhere else you hear it, with Bob it was a question and Bob wanted the answer. I'm telling you. Bob changed lives.

Bob also managed to retire before he was fifty to his own island life just off the coast of Seattle, so it wasn't just morality that he was right about.

As for me, somehow I got a job offer to teach at a college in rural Texas and was so convinced that I'd never work in another restaurant I left my good Doc Marten work shoes and my wine key in my locker since I'd never need either of them again.

It would take a few years but, suffice it to say, I was wrong.

Chapter Eleven

THE TASTE OF METAL

BEFORE MY FIRST DAY AS A SERVER AND OUR FIRST SATURDAY as a crew is about to begin, Dan finishes up pre-shift details about sections and which wait assist is with whom, and he says a new guy, Flaco (to replace Nikki) is with me and Claire.

"Get to know each other," Dan says, "and make sure everybody knows what to expect from you. If you want the wait assist to drop silverware for the next course, fine, just tell them what goes to what position. If you want to do it yourself, fine, but you need to do it that way at every table, all night, make sense?"

We all mutely nod, and I nod to Flaco, who appears a little out of breath, almost as though he just ran here. He appears to be maybe ten years older than me, and though he's certainly not at all overweight, there's nothing particularly skinny about him either and I wonder where his nickname came from.

"Chef?" Dan says, addressing Jorge's back. "Got anything?"

Jorge stirs like some kind of recalcitrant mastodon who can't be bothered with little things or little people. He glances, barely, over his shoulder as though he hadn't been paying attention at all and is surprised to see us.

"Yeah," he says finally, rubbing the big rooster on his forearm. Most fine dining chefs wear their fanciest chef's coats for service—the stark white ones with those double rows of satin buttons like lights on a landing strip, usually with their name embroidered in cursive right above their heart. Hubert even had the German flag embroidered on his collar. Not Jorge. He wears the same thing all the rest of the cooks upstairs and down wear: just the simple, short-sleeved button-down shirt like every prep cook and dishwasher in every greasy spoon in the country. Some might take it as disrespect toward his position or a way to cockily say, "You expect me to dress up for this shit?" But I take it as a show of humility and solidarity—and also a way to show off all of his tattoos.

"I didn't get much constructive feedback from anybody yesterday," Jorge says, looking around, a bit of hurt in his voice.

Everybody is stock-still and wide-eyed. One lion. Many of us gazelles. It's a pretty uncommon situation for somebody to be harsh about you not criticizing them.

"Except Esme," Jorge says. "She said somebody said something about the compressed rhubarb with the Cynic mousse being 'fibrous' or some shit and maybe we should serve it with a steak knife." Jorge continues, his tone fluctuating between mockery and chilly indifference, "What the fuck ever. It's fucking rhubarb. Of course it's fucking fibrous. It's like with the rabbit pasta and them asking for spoons to twirl it or whatever. Fuck that shit. People are such pussies."

It's literally 4:59 and there's already a line of people queuing up outside the still-locked door, but it doesn't feel like we're about to round the corner into a cheery resolution anytime soon.

"Anyway," Jorge says, "it's Saturday so we're going to be full with all the bridge-and-tunnel assholes. Don't let them fuck up my shared plates concept. Explain, people, that it's not family style and it's not tapas—it's more like they should get a bunch of plates to share, but don't describe them as small plates. I mean, yes, they're small-er plates—but—I don't fucking know—you guys use your words and shit."

A solid beat goes by and I swear we can all hear the clock strike five.

"You heard the man," says Dan, cheery and more than a little sardonic, but not enough so that you could accuse him of anything other than agreeing with what has just been said. "Places, people! It's showtime!"

Before I can turn to go, I feel a tug at my sleeve as though my son is asking me to get him more ice cream. It's Flaco, and he's nodding urgently toward the back.

"You are Matt," he says, pleading, but I don't follow for a solid couple of seconds until I finally realize he's saying my name. "Matt!" he says again and goes ahead and takes the lead, still pulling my sleeve, into the back.

"This," he says, pointing to half of a cheeseburger sitting on one of the dishracks.

I haven't said a word to him yet and still don't know what to say. I wonder if he found it there when he came in or if it was in the elevator or something and he's worried that he'll get in trouble if he throws it out or I just have no idea.

"Yeah?" I say.

"You," he says. "Matt. Eat."

"Oh," I say, stunned. "Thank you, thank you. Gracias, Flaco."

"Okay," he says, and he pats my shoulder as I dutifully pick up the burger and take a bite.

He grabs a fistful of unpolished knives and begins to shine the water spots off them at a rate that I'm worried will rub the silver from the flatware, but I'm too impressed with his speed and also the tremendous flavor of the burger. It must have come from downstairs—I guess he knows people already despite the fact that it's his first day.

"Come on, Matt," Flaco says, leading me back out. "The rock and the roll!"

"You got it," I say, pretty sure I won't be able to keep up with him despite the fact that technically he's my wait assist.

It's kind of a complicated thing to be a server and have a wait assistant. If you don't know what you're seeing, you might think that one guy—the lead server—is just talking and loafing while the other guy—the wait assist or WA, pronounced, awkwardly, "wah"—is doing all of the work, making sure the water glasses are always full, the napkins folded, the silverware clean, the table free of any crumbs or spills, and, once everybody at the table is finished eating, all the plates cleared.

It is a labor-intensive position, the WA, but it also comes with a tremendous bonus: you basically don't have to talk at all. You don't even really need to know English or whatever language is spoken at a given restaurant. You just need to speak Restaurant: water all guests when they sit, remove unused glassware and silverware from empty place settings, refold their napkins when they go to the bathroom, serve from the right, clear from the left, ladies first, protein closest to the diner, veg away, spoons and/or sharp knives down before the course necessitating them, and so forth, as well as the preferences of the specific server you're working with and, of course, a few basic nonverbal cues or gestures such as somebody pointing to an empty beer glass or somebody air-signing their signature to get the check.

Wait assists work their asses off, but they don't need to know anything about the food or drink that's served, and, for the most part, they don't even need to run the food—that's for, go figure, food runners, mostly. Their jobs are hard too, but they're also limited in scope. Sure, they have to announce what they're dropping off and be prepared to answer questions about the dishes, but if a guest has any question or concern or anything, both the WAs and the food runners can just say, "I'll send your server right over."

Wait assists and food runners are basically a fine dining restaurant's answer to the more modest busboy position, and no matter how high I ever climb on the ladder of restaurants, I will always, I hope, have the soul of a busboy. Of course, it's where I started in restaurants at Miss Katie's and again at Thornton's, and, technically, it was basically the first thing I did here at The Brewer's Table last night, but it's also how I prefer to go through life. Kind of like Jorge and his dismissal of the formal code of dress and conduct, as both a server and a bartender and also as a teacher and a man, I try to remind myself that really, no matter my titles or degrees, I'm no better or smarter or more talented or more deserving than anybody else. Restaurants tend to be really good at reminding everybody of that, whether you believe it or not.

But, I have to respect the workflow in a place like this, and try as I might, when it comes down to it, I was hired to get to know the food and beer and be in charge of communications with guests and the bar and the kitchen, and Charles was hired to run the food, and Flaco was hired to basically play free safety and make sure everything goes according to the plan.

Thanks to Flaco, my protein level has stabilized—I didn't know it was wonky, but he seems to be something of a seer—and the night begins without too much fanfare. My section is half of the chef's counter and half of the banquette. Right after we open, Omar shows up, but he mostly just skulks around the bar as though he couldn't get a reservation and chats with Miko, and not long after, Linda appears along with her husband, Todd, and they both leave Omar to go to the end of the kitchen line to talk to Jorge.

I find it odd that Todd doesn't seem to want to hang out at the bar with Omar, but he seems more comfortable standing with his back against the big glass windows like he's at least temporarily in charge of security as Linda and Jorge go over something. He's probably fifty, maybe even sixty, but his long hair and Merlin's beard are both a silvery gray, and when coupled with his half-lidded eyes, his crossed arms propped on top of his brewer's beer belly, and his heavy-metal black hoodie, he comes across as not only the kind of guy who could kick your ass if he wanted to with minimal effort but also someone who could likely as not cast a spell that could sew your mouth shut or turn your bones into string.

Linda, on the other hand, perched now at the end of my section, leaning on seat 112's chair, is so visibly anxious it's actually a surprise to see that her fingernails aren't gnawed down to the quick. Nothing is going wrong, but, at the same time, nothing much is happening. I mean, it's only a little after five, but it is Saturday and it's still opening weekend, and despite the fact that the beer hall downstairs has been running a more than two-hour wait, we're still less than half full. Linda's gaze roams around the room from concern to concern, her mouth moving as she tells herself whatever the matter is.

I'm still waiting for my first table of the night, so instead of hiding in back and polishing more already polished silverware, I reckon it's my time to step up and introduce myself to the woman who is essentially my big boss.

She's a strikingly pretty woman, probably the same age as or a little younger than her husband, but she's better made up and looks less worn for the mileage. She's got long, curly blond hair and these blue metallic glasses that remind me of the ones Gary Louris of the Jayhawks wears. I think of leading with that as I approach her but then remember some stray bit of trivia from orientation that she and Todd travel the world in

search of obscure heavy metal bands and decide not to invoke a Saint
Paul folk-pop duo as my entrée.

But it doesn't matter.

It doesn't matter because (a) I have already failed to impress the
authorities at Surly with my musical prowess—I still have yet to hear
even one song that I recommended—and (b) she's still talking to herself when I sidle up to her at the chef's counter.

"Sixty thousand dollars," she muses, staring up at the ceiling now.
"Sixty thousand fucking dollars."

"What was?" I try. I make sure to not commit all the way to leaning
on the counter with her in case she'll be shocked or offended by my in-
terruption.

"Those pieces of . . . of cloth. That fabric," she says, nodding upward
again at the ceiling. It's finally then that I gather that she means the
dark-red banners hanging horizontally from the ceiling. They don't really look like anything or attract attention. They're just like short cur-
tains running the length of the restaurant every five feet or so. "They
said it would be too noisy with the open kitchen and all the hardwood
and the big glass windows, and so for sixty fucking thousand dollars,
they hung these . . . drapes from the ceiling."

"Wow," I say, preparing to motor on.

"Wow is right," Linda says, and she finally looks at me, and it is the
stressed-out, crazy-eyed look of someone on the brink. I swear I can
see the blood vessels in her eyeballs pulse with tension. "They had better well fucking work. Do they work?"

"I know," I say. "I mean I don't know. They were there before I got
here so I haven't ever known any different."

Linda sighs through her gritted teeth, and I feel like I should give her
a hug or something, but I'm again afraid of breaking some kind of professional intimacy law so I just apologize and pretend that something
urgent requires my immediate presence on the other side of the room.

Pretty soon the restaurant finally starts to fill up. This early in the night,
I had expected to see what my Irish friend Jim called the Blue Rinse Brigade, but rather than a geriatric assembly of folks with canes and porta-

ble oxygen tanks, the restaurant feels like an extension of the brewery. Around our walnut tables glittering with ridiculously polished flatware and spotless water glasses are more dudes with heavy beards and hoodies than a Magic: The Gathering convention. Except these guys, their beards, and also their girlfriends are, if one can gather by the color and decoration of their garb, like Todd and Linda, all about metal.

It seems to be true of much of the brewery industry, as though there's something heavy metal about brewing beer. What is it, other than the need for comfortable, steel-toed shoes that serve one well whether in a mosh pit or in a brewery, that the two have in common? A proclivity toward the outer rungs of society? The desire to name your own price and sing along to a tune of your own making? Being misunderstood? Not wanting to conform? A predilection toward power chords? Loneliness?

Maybe I'm onto something here, but one thing I think is ironic about brewers being like heavy metal guys is that, sure, brewers love to drink beer and claim to live the rock and roll lifestyle, but they're actually highly trained scientists who get to work brutally early and stay for long, long hours for reasons I don't entirely understand. Todd and Ben and Perry and all the other brewers that I've met over the last few weeks have always been at work before I arrived, and most seem to stick around well into the beginning of our evening service. A Herculean task if they're out all night at metal shows slamming ewe blood or Glühwein from a ram's horn.

Whatever is true or false or in between, I find it beautiful and inspiring that our fussy, uptight, and meticulously polished dining room is full not of wealthy retirees looking for early bird specials but rather of a bunch of seemingly hard-core dudes and dudettes donning black hoodies with dragons, wraiths, spiderwebs, and bloody dagger illustrations and band names like Goat Whore, and Horror Toilet emblazoned between their shoulder blades.

I know, I know, it doesn't sound like the ideal crowd to be serving fancy food to, but with just a glance around the dining room, despite their silver spiked studded belts, their neck tattoos, and their long black and/or bleached hair, what I'm reading is not a scene about to burst with violence and hate and white nationalism but rather one fizzing with the excitement that two of the most misunderstood and most often hastily stereotyped groups—chefs and brewers—are finally united toward something greater and more sophisticated than either of their palates was supposed to allow.

Great beer. Great food.

And I don't mean Pabst Blue Ribbon and a burger or Miller High Life and some tacos—nothing against any of the above, of course; they're just staples rather than anything we'd want to call exceptional, despite the beers' bombastic names. I mean the crazy-good food these guys have been fine-tuning for months and the intensely flavorful and thought-provoking beers the brewers have been working at for even longer.

The big question, of course, is, Who else, if anybody, will care?

The night for me begins innocuously enough, and my first guests are indeed enthusiastic and gracious heavy metal folks whose only disappointment was that we didn't have our imperial stouts on tap yet. One guy with a forked and braided beard nods for me to come closer as I clear his empty glass.

"So that's obviously Jorge," he says, gesturing toward Chef. "Who else is here? Anybody else I'd know?"

I take a closer look at the line and recognize with more than a little embarrassment that, other than Jorge and, kind of, Dustin, I don't know who anybody is. I mean, I know the name of Pat, who almost beat me up for not yelling corner last week, but I don't even know what her actual job is. I need Remi from *Ratatouille* to give me the rundown on who the chef de parti and the chef de cuisine and all that are.

"We're still all getting to know one another," I stammer out, and muster some urgency to get the one empty glass I'm holding to the bar.

Why haven't we all been introduced? I wonder. Now I'm curious.

For reasons I don't entirely understand, there are the cooks who work on the line—at The Brewer's Table, this means out front before a live studio audience—and the cooks who work in the back, tucked into a corner of dishland by all the racks of unpolished silverware and glasses and plates and dishes, not to mention the shelves and shelves of Jorge's and Dustin's science experiments, such as yearlong-fermenting strawberries, racks of sprouted grains, and nearly a dozen different kinds of vinegars they're brewing.

The cooks on the line are always busy chopping and sautéing and braising and broiling and just otherwise always doing cool shit with knives and fire. The cooks in the back, I imagine, are in charge of all of the unglorious, tedious, super repetitive work such as portioning out cuts of pork jowl by weight, tearing up dozens of loaves of bread every day for our panzenella, Frenching rabbit legs, preparing the gallons of the marinade for the octopus, Cryovacking all of the veg and fruit for the "compressed" components on our dishes, and any other number of jobs for which they'll never get any recognition.

After I greet a new table at the banquette, I go downstairs for another sani bucket, and when I catch the just closing elevator to go back up, inside is Maggie, one of the prep cooks, with an entire rack of the squid ink tagliatelle to go with the rabbit.

"That's a lot of pasta," I say stupidly in recognition of the obvious fact that Maggie has a cart laden with six sheet trays of pasta in front of her.

"Oh my God," she says, "you have no idea. I've had to roll all this shit out for tonight and when I try to do it upstairs so close to the dish pit, all of the humidity totally fucks with the consistency so I had to schlep everything downstairs and do it in pastryland."

Maggie's got a kind, generous smile and the sensitive, almost hurt eyes of a river otter. She's probably five or even ten years younger than me but seems to have spent most of her adult life working in restaurants and accordingly has tired-looking shoulders.

"That," I say as the door begins to open up to our second floor, "sounds like a shitload of work."

Maggie smiles her sweet smile, grateful, it seems, for somebody to briefly talk with.

I gesture for her to go first out of the elevator, but she does the same, and we both nearly get caught in there again as the door begins to close on our showdown of politeness.

As the night goes on, I wait on a few brewers at the counter who are briefly joined by Todd. He shares a beer with them and then leaves them to their meals, which they are all so totally into they don't even look up until their plates and glasses are completely empty, and then there's a four-top of nice but competitively erudite people who keep trying to make witticisms of our menu items ("Ha-ha-ha!" one of them guffaws, "that's what 'hamachi said!'"), and then a quiet guy in a backward baseball cap and his enthusiastically kind girlfriend are sat in my section at my last open table. They sheepishly order a couple of the snacks, and before long the guy is on his third Todd the Axe Man—a pretty high ABV beer that Todd made in collaboration with Amager, a Danish brewery that insisted the beer be named after him for his scorching heavy metal guitar chops rather than his forestry skills.

And then, in a surprise move, Maggie sits down with them.

"Hey," she says to me, clearly not needing to elaborate. She orders our lager and a couple more dishes for them to enjoy as a table, and knowing I don't need to elaborate, I leave them to enjoy each other's company. And though it's always a little weird to wait on people you work with, I can't help but love the way that Maggie's face takes on an effusive glow as she tucks into the dishes, and then her second beer, and, for once, gets to tell people the tale of how everything is miraculously made.

It's not arrogance, not on Maggie's or anybody's part, that's for sure. Food and beer this idea- and labor-intensive don't happen as a result of any one man's or woman's work—that much is becoming more and more clear. What we're doing here, whether or not it ends up becoming popular or successful, is yet to be seen.

But what's astonishing to me about what Maggie's doing is that she's getting to both make *and* enjoy it. I can't think of a single time in any of the other restaurants I've ever worked in where any cook other than the executive chef gets to do or say anything in the dining room, never mind eat there. If Jorge's letting—or, heck, maybe even inviting—his cooks dine in his restaurant on this, the opening weekend, maybe he's a softer guy than appearances allow.

Similarly, Dan mentioned to everyone that after we get things going for a couple of weeks, he wants us all, with a guest of our own, to come in on a day off and sample everything—on the house too. This blows my mind. I mean, we've already gotten to try everything on the menu, albeit sharing one dish with ten or so other people. But to get to do so in a civilized way, one plate at a time, with a sweetie, for free? Crazy. This should be normal but, I assure you, it is not. Especially among back-of-house staff, prep cooks and line cooks and dishwashers can work for *years* in a restaurant never having tasted a single dish.

After Maggie and her friends leave, so does my other table. I assumed they were the "bridge and tunnel" people Jorge warned us about during pre-shift, but after they leave Linda reemerges and comes up to me.

"Thanks for taking care of the architect," she says. She doesn't smile or frown or otherwise emote in a way I can read. I bet she's a tremendous poker player.

"The architect?" I say, wondering just what the hell this insidious nickname is for. Maybe some mob cleanup guy or an accountant who's not afraid of creative mathematics, thinking back to my days at Miss Katie's.

"The architect," Linda says, "as in, of the building. The same guy who did the Walker."

My initial thought is that she's making another obscure reference, this time to the Chuck Norris classic TV drama, *Walker: Texas Ranger*, but she's not. She means The Walker. The museum. Minneapolis's world-renowned contemporary art museum. Duh. I knew this. I had forgotten this.

"Oh," I say. "Um, you're welcome?"

Some opening weekend, I think. In the same section, I have waited on everyone from the head brewer's friends, to a meek and mild prep cook, to the actual guy who designed the entire space we now occupy.

I'm sure as hell glad they didn't tell me in advance, or I probably would've OD'd on Sofia's vitamins.

From here, things progress but without much drama one way or the other. At both Surly, which is busy but civilized, and the lakeside place, which is busy but always suicidally frantic, I work seventeen out of the next nineteen days and find myself becoming a stranger in my own home. Given that I'm working until midnight or more almost every night of the week, I'm a zombie in the morning until I can stumble my way to a nap, and when I can't, I'm churlish and ugly.

One weekend day after getting less than a handful of hours of sleep, Jenae and Emory and I go out for brunch in Northeast Minneapolis, but we're waylaid by the Art-a-Whirl crowds as the streets and most businesses are overrun by self-proclaimed artists and collectors of every stripe. Scarf and beret wearers are everywhere and they're by and large drunk. And given that I now work most of my nights for these people, I'm less than thrilled to be spending my off-hours with them.

We finally almost make it to the Anchor, our favorite fish and chips place, but as we make to cross the street, Emory hops ahead of us into the road. And though he didn't just leap in front of a UPS truck or something, there's something about it that feels like a major, potentially life-threatening move, and I'm sure my entire face is full of disappointment and blame for him and, by implication, Jenae too. Because now that I'm working all of these extra hours, who the hell is taking care of our boy and teaching him to behave like this?

I don't say or really believe any of this, of course, but I'm sure I arch my eyebrows in as accusatory a way as possible, and both Jenae and Emory look at me and then at each other as though to say, Who the fuck does this guy think he is? and then walk in to the restaurant without me, leaving me stupid and skulking in their wake among the up-and-coming mimes and yarn darners of greater Minneapolis.

The days and nights go on, and, initially, they get worse. Dan seems disappointed with our mastery of the menu, and so before our next Saturday shift, he surprises us with photographs of every single dish. One by one, we have to meet with him in the private dining room, and we're not allowed to leave or work until we can name every single ingredient on however many dishes he wants to show us.

This does not feel like an academic drill. This is the same guy who sent somebody home for wearing black rather than dark-blue jeans.

Sofia is called first and, twenty minutes later, staggers out as though her blood has been drained by leeches.

After she limps by, I accidentally look up, and Dan catches my glance and gestures me wordlessly in the boardroom in the same way I once saw a Massachusetts state trooper standing on the side of the road pull over a car just by pointing at him and then the curb.

He has me sit opposite him at the big table and begins without a word.

First he shows me the panzenella—our take on the traditional Tuscan bread-and-tomato salad—and I manage to name everything except one.

"The white thing," I stammer. "Lardo?" I don't know what that means or if it's something we make or buy or just is, but decide to just see what the survey says.

Dan doesn't react one way or the other and moves on to the next photo. I don't know if that means I've already failed or if I've passed so far and just keep telling myself, "Head in the game, Batt! Head in the game!"

The next photo I recognize as the crudo. I get the main notes—compressed hamachi belly marinated with guajillo peppers, cilantro, and garlic; huitlacoche; chive flower; compressed avocado; compressed cucumber; spiced pepitas—and something else. I can't figure it out. Amid the mini-galaxy of multicolored dots and swirls on the plate, there's another rectangular element, and all I know is that it's not a fish or a sauce or a nut or a garnish. I've got nothing.

Dan shakes his head as though this is the last strike I get before he hammers down on the EJECT button.

"Masa cracker," he says, shaking his head in a way that says "too bad."

He continues to stare at the photo and then, grinning maniacally, says, "And just what is huitlacoche?"

After the soft open and more studying, I think I've got this, but I don't want to say it. But what choice do I have?

"It's a Mexican delicacy," I smile. "An umami bomb also known as 'corn smut.'"

"That's one way to put it," Dan says and tucks the crudo photo away and shows me the cauliflower, my favorite, and I nail every ingredient—even the micro cilantro and the sprig of mint.

"Sweet," Dan says, tucking the photo back into the stack. "Have a great shift."

That's it? my expression asks, sure that he was going to make me do the long march through every dish on the menu, but he shakes his head no and out I go.

* * *

I'm feeling pretty good about myself when my first table is sat and a family of five—parents plus three tykes under the age of six—are parked in the middle of my section.

"Hi, everybody," I say. "Welcome to The Brewer's Table. My name is Matt and I'll be taking care of you this evening."

"Great," the dad says unemphatically. "Could you grab us the kids' menu?"

Blood begins to chill and run toward my feet in anticipation of what will come. Namely, the harvesting of my organs by Jorge.

"I'm so sorry for any confusion, but the menu you have is everything we'll be serving today." I know I sound like a simpering ass, but I just don't yet have down that cheeky, funny way good servers can downplay the biggest faux pas as something we can all mutually chuckle over.

"No problem," the dad says, as smoothly and confidently as someone who has brought essentially three babies to a fine dining restaurant on a Saturday. "We'll just order off the downstairs menu for them."

Now my cold blood begins to drain from my toes and pool around my feet like the train of an awful, awful dress.

Sure, I explain, it seems like geographically we're so close, our upstairs and downstairs restaurants, that we should be able to get anything from either place no problem. And my Columbus Fish Market training of the-answer-is-yes-what's-the-question really, really wants to tell them You betcha!

But we're under strict instruction not to. Acknowledging that it couldn't possibly sound more pretentious, I explain that we are a finely tuned operation where timing, plating, and steps of service are all of the utmost importance.

"As crazy as it sounds," I try, "it would be like trying to get Terry Gross to do the Marketplace show and vice versa—sure, they're both produced in the same studio, but Kai Ryssdal is no Terry Gross, so . . ." Lame analogy, even the toddlers know, staring at me with blank, hungry eyes.

"So?" the dad says, not with hostility but not without some bite. "My kids are supposed to eat—what?—hamachi collar and guinea hen?"

"Actually," I say, about to tell him that the guinea hen eats a lot like chicken (I can feel the heat of Jorge's hunting rifle laser scope on the back of my head), but the guy cuts me off.

"Please just ask the chef if there's something he can do for kids."

"Of course," I say, and mope my way to Jorge on the line, fully expecting it'll be the last time I mope anywhere.

Things are picking up, and service is brisk enough that Jorge is reading and managing tickets and calling out all-days to Dustin for at least a full minute before he turns to me. When he does so, it's not just a hey-man-what's-up? kind of casual thing. Instead it's like he's pulled the big lever in the factory with the label on it that says WARNING!!! DO NOT TOUCH UNLESS YOU WANT ALL FORWARD PROGRESS TO STOP, LIKELY AS NOT BECAUSE SOMEBODY HAS A STUPID QUESTION.

So he pulls the lever and all forward progress stops, and I stammer out something in as weak chinned a voice as I dare like "Chef, hi Chef, sorry Chef, but would it be okay, Chef, I mean I know it probably isn't, but I didn't want to assume even though I think I know the answer, though, again, I'm not trying to assume anything, but there's a family at 209 with three little ones and is there any way we can . . . I mean you . . . what would I make them? . . . ha-ha, imagine that! . . . anyway, can you make them anything for their kids?"

Jorge looks across the line at Dustin, and Dustin looks down at his *Night of the Living Dead* tattoo—or maybe he's just wiping sweat from his forehead with his shoulder—and then he shrugs. "What the fuck. Rabbit pasta without the rabbit?"

Jorge nods. Looks at me. "Rabbit pasta without the rabbit," he says. And then he smiles a cruel, death-eating smile. "*Conejo* or no, it's the same price."

I am grateful for the exception to the rule, but I expect, again, the worst when I return to the table to explain Jorge's reply, but they just nod and say, "Sounds good. We'll take three to start with, and then we'll order for ourselves after that."

Bullet dodged! I glibly think and go punch the order in, but as I see the ticket print by Jorge and hear him call it out to the line, something bitter begins to sink in.

The pasta—the very stuff Maggie was fussing with the other day—is tricky not only because of the humidity of the building but because it's made not just with egg yolks and flour, as with traditional handmade pasta, but also with squid ink. The pasta, I had forgotten to mention, is black.

Very metal, now that I think of it, but what kid wants heavy metal pasta?

Because I didn't communicate this with the family I'm more than a little panicky when I pick up the three bowls of pasta at seventeen dollars each to take to these toddlers, but Dan's there at the end of the line, already having been informed by Jorge as to what's going down.

"Tell them they're made with chocolate," Dan says, with a wink.

How is that better? I wonder and walk as purposefully as I can to the table where I will try—and fail—to convince these three little baby birds that the meal I have brought them is not just dinner but dessert too. I wish I could ask Flaco to hook me up with his hamburgesa connect so I could somehow just get these kiddos some proper kiddo food, but I know there's no way.

After a short while, Flaco will clear the mostly still full plates of pasta from the kids and take them back past Jorge and on to the dish pit and right into the trash can that stands within an arm's reach of Maggie.

I just hope she's downstairs making more or otherwise not there and won't see it go to waste.

Chapter Twelve

IMMEDIATE SEATING

EVERY DAY, WE BEGIN AGAIN.

It is, of course, pithy, true, and stupid, but in the early days of The Brewer's Table, we were reminded over and over again that, frankly, the world was not ready for us. Every day, every table, it was like we were reinventing, reexplaining ourselves one guest at a time.

One earnest-faced woman sat with her husband and another couple and, after having had some time with the menu, asked me to explain the menu.

"Sure," I said, "what items did you have questions about?"

She took off her reading glasses and shook them toward the menu.

"All of them," she said. "Tell us just what the heck all of these things are."

And there went about a half an hour of my life that I would never, ever get back.

What the world is always already for is another burger joint, a pasta place, a steak and seafood affair, a pizzeria, a taco hut. If you want to be successful in the most basic terms in the restaurant industry—for example, you want to make rather than lose money—you pretty much just need to look around for what a friend of mine in advertising once described as the "blank spaces." What she meant wasn't "find something that has never existed and nobody has ever thought to ask for it," but rather find a place where a known category of something is missing.

Thomas Boemer—that Cerberus chef who tried to help us open the lake place—noticed there wasn't really a good southern restaurant in the metro. He'd already been nominated for a James Beard award for being the best chef in the Midwest for his high-end restaurant Corner Table, but he wanted to do something different and started Revival, the Tennessee hot fried chicken place. Most Minnesota food writers and eaters alike included it among their best restaurants of the year.

He's since opened two other locations, and while I don't know anything about his profit margin or his finances, I'd say pretty much all the metrics would agree that he's been killing it since day one. But after he opened his first Revival restaurant, I haven't heard a soul mention Corner Table again (which is, alas, now closed).

Pretty much the same thing went for a really good taco restaurant—enter Hola Arepa, which started as a food truck and then became a slamming restaurant. And the same thing for a solid burger joint—enter Parlour, which, similar to Revival, opened in Minneapolis to great acclaim and then did the same thing in Saint Paul.

Make food that people understand, but make it better than they can at home, more flavorful and consistent than they can get at a fast-food chain, charge them basically whatever you want for it, come up with a solid beverage program and a decent sound system (it had better be a record player if you want the hipsters to come or, as is the case at Young Joni, an upscale pizza joint that would win chef Ann Kim her own Beard award, an actual reel-to-reel stereo), and, it would appear, all you have to worry about is running down the nib on your favorite check-cashing pen.

What Jorge didn't say was "I just closed a crazy-popular Spanish tapas joint in downtown Minneapolis after more than a decade of solid business. I should probably open a Spanish tapas place in Saint Paul, or maybe a fast-casual one in Minneapolis, or a more Mexican-influenced tapas place, or an upscale Chipotle like that white guy in Chicago." He didn't say any one of those things—especially the Rick Bayless barb (that was all me—sorry Rick!)—though they all would have made solid business sense and sense in terms of Jorge's heritage and career trajectory.

What Jorge did say was "I wonder if I can make some crazy-ass, super high-end food and pair it with beer that nobody's ever had before. And why don't I do it at a brewery. And why don't I get, for my second-in-command, that Scandinavian dude who ran the restaurant at the Swedish Institute for a bunch of years who likes to do funky stuff with homemade vinegars and pickled fruit and shit. Yeah. Let's do that."

I, of course, have no idea what he actually said, but that's more or less what the genesis of The Brewer's Table felt like. Dan said it right on the day I met him. "Duck tongues and shit." A menu with the words jowl, collar, and heart (twice!). Stuff most people didn't know was food, never mind something that they would want to order and eat.

I have it on good authority that the only way Jorge got to make The Brewer's Table manifest and got investors and Omar to back it was if he also got the beer hall food program up and running.

Though I'm sure that was challenging to begin with, by the time I was hired, Jorge had the beer hall restaurant on autopilot and never needed to go down there at all.

Upstairs, things were a little less ready to roll.

Accordingly, the first couple of months could be brutal.

We had done weeks of training and conceptualizing and visualizing and tasting and studying and prepping, and then, after the initial gust of our friends and family, nobody showed.

We were open from Wednesday until Saturday, and during those first several weeks, Dan would play "The Final Countdown" and start with four servers on the floor but then cut two or even three of them before seven o'clock, typically a restaurant's busiest time. Those week-days were pretty much run by a full kitchen—because the crazy food they were doing needed many hands on every plate—but then just one or two servers, Charles running food, no wait assists, Dan managing, and Remi or Miko manning the bar.

It was slow and excruciating, and I worked my fair share of many of those Wednesdays, but I think in large part it was thanks to Miko that we found a way to keep our chins up and marshal ourselves toward something other than a suicidally foundering start.

Because Miko—Miko was amazing.

He was one of the OG Surly brew hall bartenders. Rumor had it that he and Omar had put together a ton of the taps and plumbing and what-not, much of it frantically in the day or two leading to their opening just before Christmas. He still worked downstairs, too, but somehow Dan and Jorge had convinced him to work upstairs as well.

Though nothing about the way he dressed or presented himself represented any particular fine dining air, Miko was absolutely unflappable when we were busy, but he was also unconcerned when we were dead. He was unimpressed with anybody else's authority or money—neither Omar nor Jorge nor Dan nor anybody higher up the food chain intimidated him in any way—and he worked at the same pace no matter what the shift was like, which gave the restaurant the kind of parental stability it needed. Whatever the concern, with a glance at Miko, everybody could tell that it was going to be okay. Miko, as they say, could not give a fuck.

Neither did it concern Miko if you were young or old, straight or gay, hard-core or easygoing, lefty, righty, Democrat or—well, actually he clearly didn't swing toward yonder GOP, but he didn't care much when they happened to wander in to our little corner of Wonderland. He was a former—or maybe still current—bike messenger, and he, like me and many Surly employees, rode to work, every day, regardless of the weather, but he, unlike me, always looked suave and effortless both while he was rolling silently up to the bike rack and still when he took his helmet off and still when he changed into his dress shirt, his sleeves rolled up high on his biceps like a nonchalant model who wasn't aware that he was modeling anything. Whereas I would need a solid twenty or thirty minutes just to stop sweating, Miko would slide right from his bike to behind the bar in one effortless, uninterrupted motion. He'd tuck his far-more-authentic-looking messenger bag (a Minneapolis custom job from the fine folks at Trash Bags, I would find out much later) beneath the bar, swipe back his long black hair, and it was perfect, and then cup his other hand down over his prodigious black moustache, and it was perfect, and so it would go and so it would remain for the rest of the shift. He even looked cool at the end of the night when he was mopping while the rest of us enjoyed our shifties.

He was tall and sinewy and looked like the love child of Freddy Mercury and Prince but somehow seemed both cooler than them and also slightly over their music. I'm sure he was born *someplace*, but he was the kind of guy who, if you tried to pin him down, would end up mentioning "San Francisco, Chicago, Tokyo, some time in Brooklyn . . ." and leave so much in that ellipsis that you just knew he'd already lived everywhere you hadn't yet been able to visit. He seemed somehow exotic—maybe he was semi-feline or part wolf or who knows. Whatever he was, he was everybody's favorite bartender and coworker, and throughout the course of any given shift, we'd all strive to take a turn hanging out at the side station next to his bar, where, regardless of how busy or not we'd be, he'd always be leaning against the wall, James Dean style, with one foot tucked up against the wood, polishing glasses philosophically, looking out across the full or empty dining room like the pilot of one of Odysseus's fleet. Miko had already been from Ithaca to Troy and back, and as far as he was concerned, it wasn't all that far or impressive of a trip.

His one tell, if he really had any, was that he was slightly older and slightly more Californian than most of the rest of us, and regardless of what you'd ask Miko—You think it's going to be busy tonight? You think it's going to be slow tonight? Do you like that new chocolate sphere dessert thing with the gold leaf? Have a good ride to work? You like those Johnny Marr reissues of The Smiths albums? Want some Swedish Fish?—it didn't matter. Miko was such an oddly stony yet persuasive force of nature that almost no matter what you asked him, he'd find a way to reply in the affirmative: "For sure."

And not the Minnesotan/Swedish chef variation all too popular up here: "Ya, for süre."

This was the straight up West Coast skate punk/valley girl deal. More like "Yeah man . . . fer shur."

What made us all feel better about both ourselves and also Miko being human was the fact that he was almost entirely unaware of this little speech curiosity. Get him going during a slow shift while he was in his glass polishing trance, and you could lure him into the most gentle of traps.

You'd start by asking him a question about food, then veer toward bikes or beer or music, and then spring the unexpected on him.

"Hey, Miko," we'd say, "did you try those new Castlevetrano olives they added to the mix today? They're pretty rad, right?"

"Yeah, for sure."

Let a beat or two go by, pretend to enter stuff into the computer.

"Oh and hey, did you hear that Weezer is going to reissue Pinkerton? As if anybody's been waiting around for that!"

"Oh fuck, for sure!"

And while he's distracted by putting a dirty rack of glasses into the dishwasher, spring the last one on him.

"And I heard only posers from Reseda skate at Venice Beach anymore, but their stupid valley accents always give them away."

A solid beat goes by as Miko hits the start button on the dishwasher and the steam injectors begin to blast away.

"Fuck you, man," he says, smiling secretly as though he could hide it beneath his moustache.

A funny thing began to happen around this time. Given that we were a new restaurant in arguably a most high-profile location thanks to the incredibly booming business of the beer hall below, I fully expected that Rick Nelson, the food writer for the *Star Tribune*, or Stephanie March, from *Mpls.St.Paul Magazine*, or Jason DeRusha, the excitable and generous food enthusiast from WCCO news who had his own food blog—I mean, I thought *somebody* would come by and give us a review that would put us on the map.

But they didn't.

Months passed and we were well on our way into the second half of summer and not a single reputable food writer made a visit so far as we knew.

"Dan," I said one day before our shift started. "What gives, man? Where are the reviews?"

"Oh," he said, typing away at whatever spreadsheet he was always working on at his laptop. "I wouldn't sweat it. They're just giving us some room to breathe."

"Really?" I said. "Isn't a solid review just what we need to get things rolling?"

Dan kept typing, and I started to turn toward proofing the next table when he finally replied.

"Anybody who's worth worrying about knows that they need to let us get our shit together. Who the hell really expects a restaurant to open one day and be totally kick-ass the next? It takes time."

I nodded in agreement, but I was surprised if not actually shocked. I had thought all of the training and prep and memorization and testing before and immediately after we opened was so that we would totally hit the ground running. I fully expected us to open the doors and the world would take note.

But it didn't. The world, that is. People did come by, to be sure, but it was a good long while before they were consistently the people we had hoped for.

Every restaurant depends to a goodly extent on its regulars. Given that we were new and the beer hall below us was not all that much older, we could hardly expect anybody to start coming with any meaningful regularity, but within only the first couple of months, there he was, our unwitting, somewhat unwelcome first regular.

Roland.

Roland was an older guy, probably midfifties or more, without any question single, and he had a way about him that was borderline predatory, but more likely he was just a harmless bibliophile. I say the first bit because he did whatever he could to get seated in one of our female server's sections, preferably, much to her withering chagrin, Claire's. Despite his borderline stalking, he never had much to say and usually just buried his face in whatever book he brought.

Really, who could blame a single diner for wanting the momentary company of a fetching server and a distracting something to read? He wasn't, so far as we knew, the kind of guy to make lewd, suggestive comments at any rate. And his presence here demonstrated something in his value system. I mean, he could just as easily have gone to Twin Peaks or Hooters or just a straight up strip club like Saint Paul's very own—and very grim, I'm told—Lamplighter, but he came, replete with his little Dora the Explorer-esque mini backpack, here.

He'd typically show up about twenty minutes before we were going to close, and he'd first swan by the big picture window, marching with swift purpose toward the north side of the building where there was only private event space, pretending not to take a few peeks in on his way to see who was working and in which section, and then he'd march back past our window and door, and then back he'd come in, scowling and diffident as an underpaid building inspector.

But instead of demanding to see our permits or our fire extinguishers, he'd insist on seeing our vintage reserve list. This was a list of the large, 750-milliliter bottles of special brews that Omar and Todd and crew had made to celebrate either their accruing anniversaries or another year of Darkness, our perennially brewed Russian imperial stout. The prior were always different and were named, simply, after the anniversary to be celebrated, albeit with a slightly goofy, vaguely heavy metal twist on the spelling: Sÿx, for instance, an "American strong ale," or Nein, an "Imperial Smoked Hefeweizen." These beers were released in super-limited quantities; once they sold out in the first week or two, they weren't available anywhere else in the world except for private cellars of beer aficionados and here. We weren't told how many we had, but we offered them from time to time, and they decidedly separated the beer pedestrian from the connoisseur.

Of all of our guests at The Brewer's Table thus far, I think it would be safe to say that 99 percent of them had no idea that we even brewed Darkness or any of these vintage reserve beers. We'd verbally offer them to tables or show them the lists of what we had available at the time, and almost always you could see them scowl, frown, and shake their heads as though we were making ridiculous, hard-sell suggestions.

"Twenty-five dollars for a bottle of beer!?" you could almost hear them say. "Fuck that! Let's just get whatever's cheapest that has the highest ABV so we can get wasted the fastest. Unless they do Jäeger bombs. Then let's do that."

Alas. We did not.

But despite his dowdy appearance and skulky mannerisms, Roland was here for the beer. And not the cheap shit. Roland was here for the good stuff.

Once he was seated, Claire would throw up a little in her mouth and then go over to greet him.

It was something to behold. Claire approached him as though showing up for some grim civic duty, and she would situate herself on the opposite side of Roland's table, as far away as possible. And Roland would dawdle, peruse the menu, piddle-fart around as though he was going to do anything other than what he had come to always do: wear his transparent-green visor and his little Dora backpack and sit there contemplating the menu when all he ever really wanted was one of the vintage reserve bottles and—if he was feeling plucky—also something to eat. Never an entrée or even one of our smaller plates. Always just a dessert or a loaf of our bread. Because everybody knows there's nothing that goes with a huge bottle of strong beer like an entire loaf of bread.

Before too very long, however, Roland was as much ours as we were his. I mean, Claire wouldn't say so, but he was more or less harmless and never did anything more awkward or problematic than reading—I kid you not—*The Playboy Wine Guide* or *The Manual: What Women Want and How to Give It to Them*.

But it wasn't like he was asking for feedback or participation. He was just a character, and he was our character, and before too long he came in a couple of times a month.

At least he knew what he was getting into.

The usual typical disaster scenario happened on Fridays or Saturdays. And this, incidentally, is why chefs and restaurants in general tend to hate the weekend "bridge and tunnel" people and adore their week-night diners.

The weekenders, they go out because they read some starry review on Yelp or there was somebody in their How to Become Your Own Life Coach group that said there was this new restaurant and you just *had* to go. Why they can't go during the week is a good question, but they end up always rimming around the drain that is the weekend.

During the week, it tends to be only the real professional-type diners who go out to a nice restaurant on, say, a Wednesday. And business people with their no-limit black American Express cards. They can be as rude and douchey as they want—their plastic doesn't tend to bounce, and since they're not personally footing the bill, they don't wince at going for the fullness of the experience.

When it comes to table size, one is clearly the most awkward number, but Aimee Mann is also right. Two can be as sad as one.

I know it's as old as the invention of the restaurant that where to go on a date means de rigueur that you'll go out to dinner.

But going to a movie or the opera or on a hike or just about anything that means it's just you two—now that's a date with some quality time. When your date involves dinner, then all of a sudden you're opening the door of your intimacy to strangers like me, and even if you're mostly staring deeply into each other's eyes, you still have to take occasional breaks to look at me or Charles or Flaco, and, well, it's just a little weird when you think about it.

Three is not so good either, if there's a couple and then their mutual plus-one, but it's not so bad if it's just three friends. That can be pretty great because you, as their server, are usually the most welcome at a table like that—almost like you're auditioning to become the fourth member of their party.

Four, of course, is ideal, because that usually means two couples who already like each other but don't get together as often as they want, and so they're all mutually up for entertaining one another and enjoying some food along the way, but they aren't looking to us servers to actually complete any part of their friend or family group.

More than four—especially six or more—is a mistake. It's just too goddamned many people to try to be social with at a table. Have a cookout. Go bowling. Try out that new axe-throwing place. Go miniature golfing. Hell. Stay home in your rec room with all of those overstuffed bean bags and just order some pizza. I guarantee you'll have a better time than if you insist on dining out with your party of six or more.

It's always true of large parties. There's always somebody who's gone gluten free recently and feels personally assaulted by the all-you-can-eat breadsticks, and then somebody else who didn't want Mediterranean food despite the fact that pretty much every other Western cuisine stems from it, and then somebody else who had a big lunch and is preemptively pissed that he's going to get stuck with an equal portion of the bill despite the fact that he's just going to have a bowl of olives.

Even Jesus—at his final meal on earth—even Jesus probably had a shitty time at his own last supper with his large party. It seemed like such a good idea—Let's go out, you guys! But everybody wants his attention, and, as with anybody who is also part of a trinity, he's always trying to both listen and talk as not one but three, and then there are the other twelve people at the table, all of whom are angling for a little more face time with Jesus. But Jesus, sadly, knows how this party is going to end, and it isn't going to be tragic because of how long it takes the server to split up the checks or because of the confusion over the automatic gratuity.

And when do we tend to do stuff like this? Easter, of course. We, as a dumb dining culture, have practically made it law that Mom will get the day off from cooking, and after church, we all do the reverse of Jesus's day and go out as a large party of six or more for what will be, at best, a mediocre brunch and celebrate both her and Jesus's respective sacrifices with speciously warm shrimp cocktail, turgid scrambled eggs from a chafing dish, and "ham steaks" that have come straight from a can.

One such blessing happened my own way on a Saturday in July. It was one of those perfect bluebird days when the weather was so ideal that many of our reservations were no-shows because we only had a handful of tables we could fit on our patio and most folks sensibly decided to just grill at home. Even so, the beer hall and the sprawling beer garden downstairs were so crowded all day and night long that they were continuously running a more than three-hour-long wait, but we were wide open upstairs.

We'd been trying to get along with the staff downstairs—mostly by having beers with them after work; reminding them that Miko, our mutual muse, worked both in the beer hall and at The Brewer's Table; and, most important, by never using phrases like "the staff downstairs"— and they did their best to try to throw some business our way.

A big party would show up down there and announce themselves, jovially, that they were a party of six, and then they'd hear the three-hour wait time and their faces would drop.

Then, the host, trying to throw both them and us a proverbial bone, would say, "But, hey, if you're interested in something a little more, um, fine, like, dining . . . I mean, you know, not fancy . . . but, like, fine-er . . . but, you know, you can still wear your flip-flops and your culottes and shit. Anyway, they've got immediate seating available."

And hence, pre-grumpy because they didn't get to eat where they wanted to, these six march upstairs determined to see if they can't still have a good time. They are shown into my section, eyeing jealously the patio tables outside that were all already full—two strikes now against us—and they get out their reading glasses that they pretend are ironic but that all of them really need and then they get out their cell phones so they can use their flashlight apps that they pretend are ironic but that they all really need and then they see what's on our menu, not understanding or wanting any of it, and then, like Tanner did at the very first table I waited on here, the worst sign of them all, they flip the menu over to see what else we have.

And.

There.

Is.

Nothing.

There.

And then everything shifts. That is the third strike. It won't matter how cheery or helpful or instructive or sarcastic or bossy or whatever I will be. All bets are off at this point. They are stuck at a restaurant they didn't choose but can't get out of without a lot of work, embarrassment, and, of course, hunger and thirst, and so they stay. But they only stay with the unspoken but mutually understood hive mind that they can somehow turn this all around in their favor, and the best way to do this, alas, is to mock us.

It started out innocuously enough with me greeting everybody, telling them what my name was, giving them the brief lay of the land in hopes that they'd just pick a bunch of stuff in order to expedite things and minimize their contact with me. Not so much.

"Oh, hi, Mark," the self-appointed host began. "Thank you so much. But let's skip the formalities and go ahead with the drink order, shall we? I'll have an extra-dirty martini, stat!"

"Oh! Hey, wow," I say, "I hate to interrupt, but you know, being where and what we are—which is to say, a brewery in Minnesota—we can only serve the beer we brew."

I know this is not the law in the entire state, never mind the region or nation, but it doesn't change the fact that it is a law here and now.

"Fine," the host says. "Fine. I'll have a cabernet."

Like a dental hygienist not wanting to be the bearer of bad news but needing to nonetheless give the patient the news that what ensues will involve a lot of scraping against enamel and bone, I use the analogy of how when one goes to a winery, you don't expect them to serve beer. "Well," I say, still trying to be a little funny, "here we are in a brewery!"

"Babe. Are you listening?" the host's sweetie says. "Marco here is saying they don't have booze or wine."

"I know. I *said* I just wanted a cabernet. Sigh."

"Babe. Come on."

"I said sigh."

As the host and her date silently fight, somebody else pipes up.

"Hey, man—Matt, I think you said your name was?" a gentle, reasonable voice from seat 4 at the far side of the table calls. "Would you please give us a couple of minutes? I'll see if I can't get everybody on board and whatnot. Okay?" He has kind, sympathetic eyes that seem to suggest he lost a bet or didn't sign the prenup or otherwise knows it's more than just this dinner he can't get out of but he's trying to make the best of it.

"That sounds amazing," I say. I hustle over to the bar side station for moral support and any and all available chakras my colleagues might have to spare.

I look up and Miko is standing there, polishing glasses in his monastic way, nodding as though I'd already asked him one of life's big questions.

"Yup," he says, "Tommy motherfucking Bahama." All of the guys at the table were wearing them. "The new tuxedo of the exurbs. Always casual. Always classy."

"Do you think I would die a clean and easy death," I ask, staring into the dark, black, blank face of the sleeping Micros computer, "if I hurled myself headlong off of the balcony? I mean, how high is high enough?"

Miko calmly places his perfectly polished glass among the raft of its brothers and sisters on the back bar and then grabs another damp glass from the rack. "Not high enough, amigo. Not high enough."

Esme comes over from her reasonable clutch of two four-tops and a deuce.

"Oh, Matty," she says, patting my forearm. "We love you please don't die."

"Thanks, Esme," I say and step aside, not yet having any order to put in, full of fear and trembling as to what variety of hell it will look like when I do.

She looks at my table and then, by comparison, at her section of easygoing brewers and their buddies and an enthusiastic couple on a date. "I believe that your table, technically, is the worst."

"I believe," I say, "you are correct."

After some floundering and a brief but profound enough crisis of faith/identity, I decide to suck it up and go back to my table because the hangman is not going to come to me but awaits nonetheless.

I have also decided—thanks to all my training in cheesy but purportedly inspiring self-help books that I may or may not have read such as *Who Moved My Cheese?* and *Swimming with the Sharks* and, thanks to my mother-in-law, *Screaming to Be Heard* and *Zig Ziglar on Selling*—to take control of the conversation and lead myself to the destiny I deserve.

"So," I start strongly, "hey again. How are we coming on that drink order?"

"Sweet," the host's date says. "Coupla pitchers of whatever's lightest."

"Getting warmer," I mumble to myself, yet another hurdle popping up. "We don't actually serve pitchers up here, so shall we just go with six glasses of Hell?"

I should've just said "light beer," but I can't help what it's actually called. I hope they haven't looked around too carefully at any of the other tables, though. Because while we legitimately don't sell pitchers here, we do sell small, four-ounce samples of any of our beers. If they catch wind of this, well, the looming math is both simple and crucifying with the prospect of them ordering tasting sizes of every beer we have for every one of them.

The nice guy at seat 4 gives me a thumbs-up, and I figure this is as good a sign as I am going to get, but then the host pats my arm.

"So, Markie," she says, "I don't know if anyone has mentioned this, but . . . it's my birthday. I've been a very good girl," she says, tapping me again and pretending somehow to not see me flinch. "And all this girl wants is some French fries for her dinner."

Whatever little grip I had on life and on being a server again is slipping away as I describe, yet again, one of our most basic policies about not being able to order beer hall food up here. Honestly, I'm not proud of how tersely I said it.

"Matt, hey, maybe," the apologist from seat 4 starts, "let's just get a few plates to start with. Anything you recommend for a party of our size?"

"You got it," I say decisively. "Let's start with a couple of the snack boards, a couple orders of the cauliflower, and one order of the octopus. They're all my favorites such that if you don't love any of it, it's on me. Sound good?"

Seat 4 nods approvingly, but before I can scuttle away, the host's beau gives me a sideways nod to have a private talk with him as he leans back in his chair. "Come on, man," he fake whispers, "just one order of fries too from downstairs . . ." His tone suggests that I'll be rolling in singles if I do what he says. But my vaguely strict Lutheran upbringing won't let me back down.

"I would if I could, sir," I say, smiling at the irony of regarding someone wearing a shirt with a parrot on it with the title for a knight.

I walk away as calmly as I can and go to the Micros in the farthest corner of the restaurant and shift it a little so that I can keep my back to the table as long as possible. I enter my number philosophically, then, again, with purpose and consideration, hit BEER, then let my finger hover over 6, then, again, over HELL, then, with Esme now waiting patiently behind me for the computer, finally order the food and hit FIRE.

In another minute I have their six beers trayed up and am on my way when I notice Flaco refolding the host's date's napkin and placing it on his empty seat. As I begin to set their beer down I notice the snack boards getting ready for delivery at the pass, and I get a little nervous. With individual dishes, if the person who ordered them isn't there, the food runner will return the dish to the kitchen. If the diner doesn't return soon enough, Jorge will refire the dish so as to not send out something cold. It's a righteous policy, but it does not make him happy. As it pertains to this table, I don't know if he'll do the same for apps that are going to be shared by the table, but just as Charles loads up with both snack boards and starts his approach, in walks the host's date, triumphant and smug. Charles is about to announce what all is on the snack boards when the guy plops in front of the host a little red-and-white paper basket of fries.

"Fuck," I say, not yet all the way in the server station but unable to stop myself.

Dan is there, and I do not expect him to be happy with me. "What's up?" he says, neutral and flat.

"They just," I start, swallowing hard. "Fries, Danny. Fries—from downstairs."

I know as the words escape my lips that they would seem funny and ironic anywhere else in the world. But here, in our little idealized world that we have all endeavored so tremendously to get up and spinning— here these words are catastrophic.

"Fucking hell," Dan says, and without a word goes over to Jorge on the line. I have no idea what he's going to say—I always assume everybody is as terrified of Jorge as I am, but, to his credit, Dan is not.

Still, Jorge is fixated on expoing a now somewhat busy Saturday service, but when Dan says whatever he says, Jorge looks hard at him, turns and looks at my table, looks hard at me, and then takes the grease pencil from behind his ear and slams it so hard on the counter that I can see, even from this distance, that it breaks cleanly into two.

This is it, I figure, the final strike to add to the litany of others I have racked up over the last couple of months. I'm about to get shitcanned. I take a couple of slow blinks to accept my fate with whatever will pass for equanimity when, to my surprise, Jorge isn't in my face but rather is lording over my six-top.

I hear protests of "but it's my birthday" and "come on man, nobody got hurt" and "but I've been such a good girl" that then begin to turn to "you've got to be fucking kidding me" and "I know Omar!"—but to all this Jorge just shakes his head and simply, beautifully, mercifully points to the door.

They have been 86'd.

Chapter Thirteen

DUCK, DUCK, GRAY DUCK

WORKING IN A RESTAURANT WOULD BE A DREAM IF IT WEREN'T for some of the guests, the owners, and the reviewers. Of course, it isn't that simple, because without those three elements, one doesn't need to have an advanced degree in post-Keynesian economics to understand that without diners, rich people, and food writers, restaurants don't really have any reason to exist.

Except that they do.

Without them, restaurants become priories, monasteries, places of worship but also of study, research libraries, laboratories of what's been possible and what may yet still be possible, crucibles of the conceptual and the executable. In other words, restaurants can become the food equivalent of art for art's sake spaces. This is, to be sure, the expressway to financial doom, but it is also the way to the sublime. The restaurants of our dreams.

But just as in Romantic poetry, to both diners, juries, and readers of Keats or Shelley, the sublime is both beautiful and jarring as well as terrible and scary. Just when was the last time you said to your sweetie, Should we go to that hoppin' taco joint or to that beautiful and jarring and terrible and scary new place?

The record shows quite plainly: the tacos have it.

And so, during The Brewer's Table's first few months, we saw scant few people who actually intended to be there. Even worse, among those first few who did, many were not necessarily there to wish us well or bid us bon voyage.

As I've mentioned, surprising to me but apparently simply how business is done according to the unwritten bylaws of the reputable food writers, it is considered poor form to hastily pass judgment by way of an early review of a restaurant so new it has yet to really get its sea legs. The critics had finally begun to come—I thought they'd come anonymously, but most of them booked reservations under their own names—but all of them had kept mum for those first many weeks. It didn't seem right to me, but I alone was getting panicky about it.

By then, we needed somebody to replace Rae, and Sofia had recommended her sergeant at arms from one of the last places she managed.

"You guys," she said after the usual pre-shift talk, "this is Cash. He's coming from one of those fancy-ass Loring Park places and I've worked with him there and at a bunch of other spots. He's not a typical FNG, so don't treat him like a Fucking New Guy or I'll slit your carotid arteries, okay? Kisses!"

Cash grimaced and nodded somewhat sheepishly, knowing, it seemed, that there was not much to be done about both being the FNG and everybody being told that we weren't supposed to treat him as such. I could tell from his wide, confident, but somehow still humble soldier's stance that he had been where his shoes were now all too many times. There was nothing to be done but weather the first few shifts as the random stranger in a somewhat already established crew.

And, I just realized, I was a part of that already established crew! Because half of our waitstaff came from downstairs and I had been so long out of the game, I had really been the FNG, but no more.

I wasn't yet a good server or somebody Dan would let behind the bar or Jorge behind the line or into the walk-in, but I wasn't new. They knew the ways in which they didn't really yet trust me—I couldn't change a keg downstairs or be counted upon to get more finger limes or cucamelons from the back (both somehow real things)—but they didn't yet know how they couldn't trust Cash. I know it wasn't exactly a Tony Award for Best Actor, but it at least felt like I would have been tacitly implied in the winner's "and to all the little people who helped along the way" mention.

For this and more reasons, I immediately liked Cash. First off, he was in with Sofia, which was mostly all I needed, but the somewhat shy, recalcitrant way he greeted us after her intro sealed the deal.

"Yeah, so, hey," he said after Dan motioned for him to introduce himself, "like Sofia said, I worked most recently in Minneapolis and, good God, like a million other places here and then before that in—I don't know—New York, New Orleans, San Francisco . . . guess I'm an old dog that's been around the block a time or two. Anyway, don't listen to her. I'm the FNG so let me have it. I prefer to suffer."

I could tell from almost the minute we met that Cash and I would be solid compadres, but compadres in the cranky old man way that didn't often exist in this young person's sport. I'd be the Jack Lemon to his Walter Matthau among our otherwise millennially staffed crew of wunderkinds. But everybody needs father and/or grandfather figures, and Cash and I were ready and willing to perform those roles.

Within a couple of shifts, it became evident that he and I were simpatico. I suppose it could've been the same with him or me and Levi—we were all oldish guys with salt-and-pepper beards and hair—but Levi was always aloft in his own gauzy world. Like Miko, he was at least as old as I was, if not a little older, but he behaved in a manner that seemed to suggest he existed somehow outside of time, like Yoda or Andy Warhol or Sting, but not as creepy. For starters, he had a biblical-length beard and long hair, streaked black and gray, but styled somehow to suggest that he could just as easily model for Urban Outfitters as lead a sit-in protest of corporate chicken farms or Starbucks or non-breastfeeding-friendly workplaces. Other than the fact that Levi was a "visual artist," all I knew of him was his Mensa-like ability to memorize the data of our menu.

Shortly after Cash started, he and I were trying to remember what kind of olives were in the little bowl we gave guests after they were seated. I loved these little touches—we also left guests with house-made candies or cookies or caramels. It was the kind of thing that made 11 Madison Park or the Gramercy Tavern not just famous but beloved— and I hated not knowing all of the details about things like that.

"Fuck dude," Cash said, "I have no idea. Green ones and black ones?"

"I'm pretty sure one of them was . . . Greek?"

Levi glided into the server station as though he was on a magic carpet. "Picholine," he began, having somehow sensed our feeble reaching, "Castelvetrano, kalamata, of course . . . ," he drifted off, staring into the middle distance between the tables on the patio and the grain silos beyond the beer garden. Cash and I exchanged looks wondering where

Levi had gone, but then he snapped out of it and said, "Niçoise—how could I have spaced that?!—and, of course, Cerignola."

"Cerignola!" Cash said, slugging back a shot of coffee just before the doors were opened for the beginning of service. "Just like mom used to make!"

Levi shrugged and off he floated to help seat the first rush of incoming guests.

Things had been better recently—the dishes were getting more and more dialed in, the flow of service from the kitchen was bordering on the absurd it was so smooth, and the quality of the food was just ridiculous—and yet it still didn't feel like we'd hit our stride as a restaurant yet.

One of my favorite things we'd begun doing, however, was becoming as much an albatross around my neck as a point of pride. It was called "Beer and a Bump"—a throwback to the wild eighties, I guess. I am and have long been such a square that I didn't know it meant anything other than what we did with it. For ten bucks, the idea was, you got a special concoction the chefs competed with each other for every day, along with a little beer to go with it. It was supposed to be a take on a boilermaker—a beer and a shot of booze (or cocaine, depending on the policies of your favorite watering hole)—but since we didn't serve booze (or coke) the beer was still the beer, a little four-ounce serving, and the bump was a little "shot" of food. The cooks pitched the dishes to Jorge, and whichever one he liked the best was what we served that day. That part I loved. I also loved the fact that Jorge insisted we not tell guests what the beer or the bump in question was. If they wanted to know, they had to order it.

It got a bit clunky when we had to fit this mysterious curveball in with the descriptions of the rest of the already sufficiently austere menu. Plenty of guests took this as their final and certain cue that they were in decidedly the wrong place. But, plenty of others—the guests who would become our regular and most ardent fans—were thrilled at the chance to order something so strange and ephemeral. Whatever it was going to be, these intrepid diners concurred, it was going to be special. Not "blue plate" special as in something for the masses. But special as in nobody's had it before today and nobody's going to have it again.

For reasons I never fully understood, we had an awful lot of pickled herring in those early days—Dustin must've had a connect from his time at the American Swedish Institute—and since it is a rare diner who leaps at the opportunity to actually choose to order borderline-rotten baitfish, the almost unwritten rule of Beer and a Bump was that it was going to involve said herring. We had a bump that was called "Expressions of Herring," which featured, if memory serves, a herring foam, a herring mousse, and a dehydrated herring chip, served with our Dampfbeer. We had—my favorite of pretty much all of those Beer-and-a-Bump days—a dehydrated Bloody Mary with nasturtium and a cured herring spear standing in for the pickle in a tiny shot glass–sized terrarium that we finished off with a little puya chile–infused Fiery Hell—and we even had, much to everyone's chagrin, "The Herring Situation," in other words, herring ice cream, accompanied by a nice shot of Darkness, our famous Russian imperial stout.

By this time it was getting to be midsummer, and we were finally figuring out who we were as a staff. Charles had proven unflappable and always erudite as our lead food runner. Flaco was consistently fast but a little erratic as a wait assist—we didn't have dozens of different bits of flatware for each course like some restaurants with their baby oyster fork and their grapefruit spoons and such, but it was important to Dan and Jorge that each diner get the right implement for each course and nothing more or less. To save time when we were busy, Flaco liked to just give everybody new everything—all new forks and knives—at every course. Drove Jorge nuts. Miko single-handedly made The Brewer's Table a cool place for the beer hall folks to hang out at after their shifts downstairs, which kept our street cred real. Dan still had that Alfred E. Neuman way about him, but he was even-keeled, consistent, and jovial as our manager (except for the shift after he had gotten a little too into the scene at a Limp Bizkit concert—that was the only time he seemed anything less than stoked to be at work). Really, there was nothing he wouldn't do to help the restaurant or us servers. Esme and Levi were our walking Escoffier dictionaries, but both were a little fragile when faced with too many guests at once or tables on the patio, which was blistering during the brunt of summer for the first few hours of service. Sofia was consistently crazy and sweet and voted most likely to get so involved with any given table as to threaten to elope with them on the spot. Cate was so smart and beautiful and anonymously gendered with her short shorn hair that basically everybody tried to claim her as their own BFF. And Cash was proving to be an absolute beast in the thirties and forties—the section of eight two-tops. That zone was sure doom to people like me and Sofia and Cate, who tended to get too deep with tables, but to Cash, who had a way of getting in and out with an infantryman's efficiency, it was his breadbasket, and we were always grateful for it.

I had kind of developed into a utility man, or so I like to think. On days when Sofia's meds were off or Esme's OCD was getting the better of her, the hosts and/or Dan would set them up in the fifties and sixties or the seventies or eighties—both sections had only two round four-tops and four two-tops—and more often than not you never had more than sixteen diners at a time spread out over four to six tables, and neither section was ever full when the patio was open during that first summer. Cash would be handling just as many diners, but they'd be maniacally spread out over eight tables. I'd say he regularly did about 30 percent more work than the rest of us, and since we pooled our tips, suffice it to say that everybody loved Cash.

Unless it was cool and maybe even cloudy, nobody really wanted to serve on the patio. Because it faced directly west, the sun essentially set for four or five hours, blazing straight down on all seven tables out there, and we never got even a thought of shade until after eight. You could stand with your back to the sun, but that was cruel to your guests because they'd have to face down the nuclear apocalypse just to talk to you, or you could stand facing the sun yourself, but that was no better because the light would scorch your retinas and you couldn't write down their orders or read the Micros screen for the next few minutes. In either event it was brutal, and it was also farther away from both server stations, as well as from the bar and the kitchen and the dish pit. Serving out there involved more of everything. More miles. More tables. More waiting. More negotiating the heavy doors with trays of beer and armfuls of plates. More heat.

For reasons I never fully knew—was it hazing, a collective vote of confidence, or perhaps a curiosity to see if I actually had the guts to jump?—during that first summer, I was regularly scheduled on the patio. Because of the way pooling tips worked—and because I had plenty of winter weight to shed—I mostly didn't mind. For the same reasons, I also never minded closing. It wasn't who I thought I was going to be but it's who I ended up becoming.

In most restaurants where servers keep the tips they earn, they try to work the busiest portions of the week and get out as soon as the rush is over so as to maximize their profit-to-time ratio. At Surly with our tip pool, it literally made no difference to me what my personal tips were since they just went into the pot that all of us servers collectively split based simply on how many hours we worked. Since I was an oldish husband/father and had no social life, it usually didn't matter when I got done, so I was always willing to stay late and close. As a result, I think I became known to the hosts and Dan as somebody to assign wherever, since eventually I'd end up taking over the rest of the tables after everybody else got cut.

For the most part I loved this role. It mollified Dan against his doubts that I could actually do the job, it made my fellow young servers grateful that they could get out early and hit the bars or whatnot, and it made the hosts grateful that I wouldn't ever bitch to them about which section they put me in or whatever latecomers they punted my way.

As the summer progressed, the shifts at the lake place grew more and more tedious, and my coworkers there, in turn, got more and more sick of me saying things like (apropos of any given gripe about how things were falling apart there) "Actually, at Surly . . ." But I mostly didn't mind working there, since it was located where it was—by my house by a little lake—and even if we would never get our collective shit together, we were always busy enough and there was always plenty of cash to take home.

My metric for a good day at work—in restaurants, anyway—has always been that if I can leave with a hundred bucks in my pocket it was worth my time. If I can do that four or five days a week, I'm pretty much making the same money as I do at my teaching job when I'm at full pay. And, at least so far in the summer, I was. And then some.

At the lake place I was pulling at least $150 a shift, but it always felt like my money had blood and tears and probably other bodily fluids on it to boot. But at Surly, it was more complicated. Since we pooled tips, I didn't walk out with cash but instead got paid by check every two weeks, so I didn't really know how I was doing for the first few weeks since those checks also included getting paid minimum wage for all that training. But as we started to pull up on midsummer, I began to realize that every check I got from Surly was for about a thousand bucks. Sometimes more, sometimes less, but on average, I realized (way too slowly) I was taking home—after taxes—about $250 per shift. By only working two days a week at Surly, I was making about half as much as I was by teaching full-time.

I had gone into this whole sabbatical pair of jobs as a way to try to stop the hemorrhaging that was happening with my reduced salary, but it was quickly becoming evident that, somehow, I had stumbled into a couple of jobs that had doubled my income.

I was making, I realized, as much by waiting tables four days a week as I was as a tenured associate professor.

This hit me hard.

There are lots and lots of reasons for this, but mainly it was:

1. I have moved across several states, losing friends, leaving behind family members and community ties, just so I could work in higher education.
2. My university and I don't always see eye to eye.
3. For the distinct privilege of hitching my name and reputation to said university, I had incurred just shy of a hundred thousand dollars of student loan debt, which leaves me with, to this day, nearly eight hundred dollars a month in loan payments that I'll keep paying for the rest of my life.
4. I was super lonely teaching. (Talk to teachers you know—our students are not our friends, and we're all usually too exhausted at the end of the day to hang around with each other, especially if we're married with kids.)
5. I was having a blast waiting tables.

There was no question about it: I took the job because I needed the money. I kept the job—and would keep working at Surly for years, even after my salary was back to full pay, even after I was again teaching full-time—I kept the job at The Brewer's Table because I loved it.

* * *

This did not necessarily go over well with my non-restaurant family and friends. I was loving work and could hardly wait to go—or even to study up on food and beer and whatnot when I wasn't at work—but it meant that when I was around, nearly all I wanted to talk about was, you guessed it, food and beer. And I was barely ever around. During that first summer, I was getting to spend a goodly amount of time with my kiddo—we'd wake up late, go swim at a public pool for a couple of hours, and then it'd be time for me to get ready for work and leave by quarter to four. And then I'd do my thing at work where, because it's the law of the land, I'd leave my phone in my bag in my locker and not have any communication with anybody except my present company—a pretty rare thing anymore—and then I'd get home at one or two in the morning, have a (very, very) late dinner, and crash until the next day, when I usually had to do the same thing over again.

From my perspective, I was still mostly taking care of our kiddo and saving us from any daycare or babysitting expenses and then making good money at night to compensate for what I was lacking to make at my college, so I was doing good. I was doing my job but still mostly treating my family right.

But, of course, I wasn't.

Not because I was doing anything bad or wrong or illicit or scurrilous. Heck, I didn't even ever have more than my one shifty after work, never mind floating over to Palmer's or the CC Club with the other servers after work. I was just at work, but I was at work at night, when most families are able, for a short few hours at least, to be together.

Jenae bore it like a champ, but it was something to bear nonetheless.

I got the gravy hours of childcare when I could let our kiddo veg on cartoons and then be lazy by the pool or have a grand old time tootling around on our bike rides or whatnot, and then when she got home she'd be exhausted from a full day at the office and would immediately have to take care of our dog and figure out dinner and give enough structure to their evenings so that Emory would be ready to wrap up the night when she needed to get to bed so that she could get a decent amount of sleep.

It rarely worked out that way.

Only, since I was simply incommunicado all night and got home after they were long asleep, I never really heard about it unless it was my day off.

In addition to all of that, now instead of having a cushy academic gig where both my students and I got off for basically every holiday on the calendar—religious, secular, bank, or otherwise—I had a job where I had to work whenever I was scheduled, especially if there was a holiday. Memorial Day? Super busy, had to work. Fourth of July? Super busy, all hands on deck. Labor Day, never know, got to have a full staff just in case.

I missed countless get-togethers, family evenings, movies at the drive-in, four Minnesota United soccer games, six St. Paul Saints baseball games, and was about to have to punt on a long, extended trip to Nebraska for the Fourth of July because, of course, I had to work.

"You're missing this—you're missing out," Jenae finally said as we sat at a Sunday brunch at the Bulldog in Lowertown Saint Paul. I was musing over what gave the Bloody Mary its oddly sweet, garlicky flavor when she grabbed my hand. "Us," she said, gesturing to the three of us at the table with her other hand. I had closed at Surly as usual the night before and was groggy and feeling entitled to being so, but Jenae's words shocked me. "*Us,*" she said again, comprehending my lack of comprehension. "You're missing us."

It wasn't a question, I realized, but a simple statement of fact. I could know as much as I wanted about malts and hops and ABVs and flavor profiles and the provenances of various foodstuffs and even the names of some of our purveyors and farmers, but I didn't know in any meaningful way how the summer was going for my family.

Shortly after this bombshell landed, so too did the food we ordered. Emory picked at his hot dog and tater tots and Jenae cut the Philly cheesesteak in half for us to split (in anticipation of their Nebraska trip we were all eating meat again), but I let it and the nachos I had ordered sit and grow gelid in the unseasonably chilly morning air.

I was doing what I thought I had to do. I was working not one but two extra jobs—and though I was loving those jobs, I was missing out on what really mattered. I was fucking up.

But we needed the money.

What the hell was I supposed to do?

The same thing I'd been doing all summer. We drove home and I hopped on my bike like a spoiled child and went to work.

At Surly things continued to evolve. The Beer and a Bump was still different every day but a new four-course food-and-beer pairing menu was being rolled out. Every week we had been replacing a couple of the beers and dishes on the menu, and even though this meant we had to study something almost constantly, this kind of innovation kept everybody interested and on their toes. We'd been open now for a couple of months, and we'd never served chicken (unless you count the guinea hen); steakl or basic starches like fries, mashed potatoes, or rice. If Jorge even got wind that something was likely to be more popular on the menu than its brothers and sisters, it was as good as gone.

For a very short period, we had a veal tongue taco that was served on house-made, Middle Eastern–style naan rather than the typical Mexican tortilla. And even though it was expensive at seven dollars for one little taco—and it was veal—and it was tongue—its popularity was actually a problem. When guests didn't know what to order, they saw the word "taco" and thought it was a safe bet, and some people ordered nothing else other than three or four of those, and we could all tell that the last thing Jorge wanted to be in life was simply a taco cook.

He was bitching about all the tacos to Dustin right as service began one night when Dan, eavesdropping, got that grin of his going. He whispered something to Cash, who began to punch something into the computer, and a moment later it was still quiet enough for everybody to hear the kitchen printer by Jorge erupt with its new order.

Jorge read the ticket and then spun around, his eyes flashing like molten lead, the veins in his neck throbbing, looking for Cash. "Fuck," he finally said, not finding him and turning back to Dustin on the other side of the line, "firing ten fucking tacos." He marked the ticket to remind himself that the items were on fire and then turned around again, eyebrow up, suspicious. "Who the fuck is this even for?" he said, slipping his pencil behind his ear, glancing around the dining room, empty except for one of the beer hall servers taking his shifty at the bar with Miko.

Dan started to laugh his heh-heh-heh-heh evil henchman laugh, and Jorge didn't even respond to him. "Fuck tacos," he said to Dustin. "Forever. We're 86ing tacos."

The dish to replace it, perhaps a bit ironically, was another inspired by Jorge's Yucatecan heritage. A tamale. It looked like just a simple tamale, still wrapped perfectly in its traditional corn husk, and a couple of sprigs of micro cilantro and a few red Fresno chile rings, and that's it, it seemed. But once we got to taste it, its simplicity became symphonic. It was as perfect as any taste of anything I'd ever had. It was all about the corn—there was the husk and then the masa harina, basically dried corn ground into flour, and there was also fresh corn inside, sweet and with a little bite. Once unwrapped, the tamale gave itself away as the platonic ideal of a humble, homemade, rustic Mexican dish, and yet in this manifestation it demonstrated several lifetimes' worth of refinement, inheritance, restraint, and polish.

On the first Wednesday of July—right when I was missing Emory and Jenae the most while they were hurtling toward the Fourth in Nebraska and away from me—I got to work and was milling around, absent-mindedly checking for smudges or fingerprints on our glass candle vases, when it was time for pre-shift. Dan cued up "The Final Countdown" as usual, but once we got to the kitchen pass, Jorge wasn't there. Dustin was in his place, and he was looking more Norwegianly serious and unemotional than usual. I imagined he was more grim because he'd have to talk to the whole staff, and that was decidedly not his happy place—having to use words and such in front of a crowd—but there was no telling. There might also be, I was afraid, bad news.

Just recently a relatively small outlet had run a review of our restaurant, and, suffice it to say, it was less than glowing. We were still waiting for the "real" reviews to come in, but every amateur critic with a pulse and at least one finger had already left their hatchet marks on Yelp, Open Table, Instagram, the predictable little hobby blogs, vlogs, and elsewhere. To those, I was impressed, nobody on staff really paid any attention. But this new one that came in, well, it was different. It was actually in print—rather than just online—and, accordingly, was physically present in the restaurant one day, a Thursday when I wasn't working, so when I came in two days later for Saturday's service it was still there in the prep kitchen, staring up like a bad report card among the chafing dishes and unpolished silverware. I didn't find it, though, until halfway through the night when I was looking for steak knives. I read it quickly, not wanting to make my table wait for the right cutlery, but I had to at least skim it.

I couldn't put my finger on exactly what it was about the review that irked me so much. The piece was, in fact, a double review of both our restaurant and the beer hall downstairs, and decidedly more attention was paid to them. About us both was criticism that was vaguely valid. The space was austere and industrial—*ahem, I thought, it is a brewery in an industrial park*—and it was loud—*yeah, it's super freaking busy*—and the waits were long—*duh, see point 2*—but, sure, the beer hall food, when it arrived, was satisfying and proportioned well. Okay, I thought, now let's get to The Brewer's Table.

About us, it wasn't so much what she said as what was between the lines. She thought there were too many sauces and too many items on each plate. She thought we were fussy. She used the phrase we hated to hear but rightly deserved. We were serving "tweezer food." She criticized us for Cryovacking food, bringing it up to temp in a sous vide water bath and then flashing it in a hot pan just for service. That was exactly what we were doing, and I thought it was freaking miraculous. Nothing was ever—*ever*—over- or undercooked, and it was ready within minutes of my firing any given ticket no matter how busy we were.

Then I realized what bugged me the most. I looked at the date the article was written and did some easy math tracking back to when she said she had visited us and where she sat—Wednesday last week—at a lone table on the patio right after we opened for service.

"Oh fuck," I said out loud to the steak knives I was still holding. "I waited on her."

But nobody had said a word yet, and more than a week had passed. But maybe tonight, with Jorge gone and Dustin at the helm, I feared, was when I was going to get singled out and demoted back to busboy and the short apron.

Dan began the pre-shift business as usual. He highlighted Hop-shifter, a new beer with a rotating, different hop varietal every couple of weeks or so, and let us know that at some point in the evening we'd be running out of Schadenfreude and to keep an ear out for that.

Dustin stood by mutely as Dan had us discuss various beer pairing options for the tamale. Some thought that the Hell lager was enough (the bready, corn-on-corn flavor did the trick), while others thought that Fiery Hell gave it a little something extra (the puya chiles brewed into the beer acted like a dash of hot sauce), while I thought that our Witch's Tower brought something unexpected and acted like ghost garnishes (the session brown ale had distinct notes of cardamom and mint, which made it a beer I didn't care to drink on its own but with the tamale took on a kind of Yucatan-via-the-Middle-East vibe that I just couldn't get enough of).

We could've gone on, but Dustin nudged Dan that it was time for something else.

"All right," Dustin said. "Any other questions? No? Good. Shall we move on?"

Fuck, I thought. This is where they give me the axe for the bad review. Sweat began to dot my scalp in anticipation of the fire I was about to face.

But then Dustin picks up a huge magnum of champagne he had been hiding and with his other hand hoists his chef's knife. "You guys have been doing a fucking awesome job this summer," he says, "and, as you might or might not know, we just got our first really cool review, so, you know . . . that's a big fucking relief. So, congratulations!" Many of us look around at one another, not having heard a word about any of this, but apparently there isn't time for details. It can't have been the review from last week is all I know.

Dustin begins to count out loud and takes two practice passes with his knife at the champagne's neck, and with a one, two, three, he slashes at the cork.

But instead of frothy, jubilant champagne spurting out, a shower of shattered glass sprays ecstatically all over the pass, the entire expo line, the garmo station, and who knows where else.

"Well, fuck," Dustin says. He's still smiling but sheepishly now. They suddenly have an awful lot of mise en place to pick through or redo. "Have a little glass in your bubbly everybody and get the fuck back to work."

Turns out the review in question was not one but many. Just a week after we were panned, there were astonished and epiphanic reviews from the Minneapolis *Star Tribune*, the St. Paul *Pioneer Press*, *Mpls.St.Paul Magazine*, and, probably the crowd favorite, one on WCCO's Jason DeRusha's blog, where he said he "just had one of the best meals of my year and it was on the second floor of a brewery." His and other reviews used words like "incredible" and "amazing" and "fantastic" and "Instagrammable" and "perfect."

With these reviews, Dan and Dustin and Jorge, when he returned later that evening, were relieved but mostly unmoved.

They kept doing what they had been doing and urged us to do the same. Despite the steep learning curve in those early weeks, Dan kept us hungry for more beer and food knowledge. On any given day, we would show up to our pre-shift and he'd take us on a tour of Dustin and Jorge's rooftop herb garden or have us do a blind taste test of all of our various sauces where we'd have to name the sauce and all its ingredients or, likewise, do a blind taste test of five of our beers and have to name which each was and identify the malts and hops to boot. We did tours of our brewery and got to sample in their raw forms our grains and malts, and we also got to meet the purveyors of our produce and our meats. We invited them in to dine with us, too, and give us their feedback on our food-and-beer pairings. We even got to take a trip to Au Bon Canard farms in Caledonia, Minnesota, to meet Christian and Liz Gasset, who, instead of having children, chose to raise Muscovy ducks. We've featured the ducks on our menu in various forms, from their tongues, of course, all the way to foie gras in our beet salad. They are one of only three purveyors of foie gras in the entire country, and we as a staff got to spend the night with them, camping in their pasture, where in an act of ironic beauty our hosts kindly force-fed us pound after pound of foie gras amid the constant din of quacking.

Chapter Fourteen

NO PLACE FOR PITY

A YEAR HAS NOW PASSED AND THE RESTAURANT HAS CONTINUED to evolve, and even though I'm teaching again, I'm still working at Surly on Wednesdays and Saturdays at least. More if I can. Today is one such Saturday and it's late, very late, in the morning. I worked last night and closed, which means Dan and I were the last ones out the door. You know it's a late night when the chefs have all already left and turned off the exhaust hoods over the flat top and the grill stations. They're always going, this nonstop Moroccan sirocco, blasting like a hot wind to evacuate the smoke from the back of house. Not until they're shut off do you realize how loud they'd been all night long, like when there's a power outage and your refrigerator goes silent and you have a hard time processing how you never noticed how loud they really are and how silent and kind of scary the world is without them.

One reason last night went so late was thanks to one of our new pairs of regulars. They came in, late as they always did, and they had brought a couple of their friends and also a cooler of their vintage Darkness bombers, which they shared not only with their guests but with me and Dan and Miko, too.

There are plenty of dark, syrupy, high-ABV beers out there, like Lagunitas's Russian Imperial Stout or North Coast Brewing's Old Rasputin, but Surly's Darkness had earned a perfect 100 points on BeerAdvocate's rating scale, making it not only a regional hit but an absolute world-class beer. It's released around Halloween every year and is available first at Surly's party dedicated to just to that: Darkness Days, a beer and heavy metal mini-festival at the Brooklyn Center brewery. Not only had our couple in question met at that party and got quickly engaged and married, but they also had a cellar full of Darkness and other interesting, rare beers and had found their spiritual restaurant home at The Brewer's Table.

Suffice it to say that it was impossible to turn them away or to even pretend that they were any kind of imposition (despite the fact that they tended to show up minutes before we closed and would then linger over each course so that their meals tended to take upward of three or so hours).

Thank God, I told myself again and again, that we pooled tips.

So by the time I rode my bike home and made a very late dinner for myself and wound down by watching a couple of episodes of *No Reservations*, it was wee-o'clock in the morning, and, accordingly, I didn't wake up the next day until it was time for lunch.

And so I did and so we went to Grumpy's, where Molly took good care of us, as always, and I let myself acclimate to consciousness via a cup of mediocre coffee while Jenae and Emory played a long game of pool. Emory got his usual Impossible Burger (back to being a vegan), and Jenae and I, exhausted from yo-yoing between meaty and meatless diets, split a regular burger and the chicken strips, and we played another game of pool, and then it was already two in the afternoon and I had to start getting ready for the big Saturday night shift.

I have to be there for our Saturday pre-shift meetings at four, which means I have to leave at three thirty at the latest, which means I have to shower as soon as I get home and press my shirt and make sure my jeans and apron are still relatively stain free (or will be in the dark of the night's service) and replenish all of my lost pens from the night before. I throw all of that plus my trusty Dansko clogs into my messenger bag, kiss everybody on the head, and roll down the street on my bike.

There's always been something about cycling that, no matter how harried or rough around the edges or whatever I'm feeling, I always feel better once I get on a bike. And getting to ride to work to Surly is always just about perfect.

Soon I pass the German Immersion School with their inexplicably

weird *Alle Sind Hier Herzlich Wilkommen* signs. I know it's supposed to
(and might actually) say All Are Welcome Here, but it's in freaking German, and what immigrant or refugee or nonnative speaker of English is
looking for any kind of welcoming sign in that language in Saint Paul?
Despite its cheery color scheme and intended message, what must that
look like to folks new to the United States?

After musing on that, I'm already across Lexington and most of the
way to Snelling, where I go past my favorite deli where the owner always says "Hey-o!" whenever he hears my order—two California Clubs
for me and Emory, a BLT no T for Jenae—and we talk about hockey and
sometimes he gives me free lemonades during his "happy hour," and
then it's past the four-lane unpleasantness by the state fairgrounds, but
I never mind too much because I'm halfway there and it's time to stop
at my favorite gas station, the Pump N' Munch.

I don't always stop on my way to work, but on Saturdays lately I've
made it my habit to pick up a Clif bar or one of those bison bars for myself (just in case there are too many reservations on the books and the
kitchen doesn't have time for a family meal before service) and a box of
Hot Tamales or Sour Patch Kids or some other candy to put in a little
coffee cup by the bar service station for everybody to share when our
blood sugar gets low around mid-shift and we can all get a little cranky.

It's the kind of gesture I practically live for but one that has taken
me a long time at Surly to feel like is welcome. I mean, *you* try offering
a handful of gummy bears to a bunch of supersmart beer/food geeks
who are all younger, better looking, and/or more heavy metal than you
are. It's intimidating.

It helped break the seal when Dan brought a box of Swedish Fish one
day, and then on another when we were going to be booked from open
to close, Esme brought a stack of Jack's pizzas for us all to have after service. Nobody would ever top what Sandi, one of the prep cooks, used to
do though. Once the rush was over, she'd throw together some kind of
miraculous casserole from what wasn't worth saving for the next shift
but was still edible from tonight; it was always odd, of course—like a
pork jowl, octopus, and shishito pepper stir-fry or a hamachi collar,
tater tot, and boquerones anchovy hot dish—but she always put her signature on it: a penis or a pair of boobs, or both, drawn in Cheez Whiz

across the whole sheet pan enterprise. (This might not translate to civilian or academic life, but in the restaurant, it was like a nighty-night hug-and-kiss from everybody's favorite auntie.)

Until Jenae started brewing it at home, I'd also usually swing by the Hampden Park Co-op to pick up a kombucha (they didn't have candy, and the Pump N' Munch didn't have anything fizzy other than soda). By then it was just a jog down Territorial Avenue, behind the office park on University, and then through the back forty of Surly's vast parking lot, which was crawling with slow-moving cars prowling for parking spots. But since I was on a bike, I got to snake through all that and I'd sidle right up to the ramp by the back door for the employee bike parking, always hoping but never achieving the dream of arriving at the same time as Miko.

But no matter—now it was time to head in. Tonight, Maggie is on the line in a surprise move, prepping what appears to be a hundred pounds of fava beans. She smiles and says, "Hey."

"Hey yourself," I say back.

Jorge is nowhere to be seen, but Dustin is at the middle of the line, prepping garnishes and such that I can't see, and he is clearly in control of the partially melted boombox as DeBarge is blaring out something about the rhythm of the night.

And thank God, because if it wasn't him or Pat, who mutually love all things Mariah, Whitney, and Janet, it would be EZ or one of the other throat- and face-tattooed prep cooks cranking out Nine Inch Nails or Tool, or—Jesus save me—Slipknot, sometimes looping their favorite anthem, "People = Shit," on repeat, and then we'd all be adequately stressed and tense to guarantee we'd have a hateful, teeth-gritting aneurism of a night.

Cash is already in the thirties and forties prepping his section, measuring with a length of string the distance between each table to ensure they're all uniform. Michele, one of our super new servers, is in the fifties and sixties proofing her glasses and silverware. And, sure enough, it turns out I'm on the patio so I don't bother changing because I'll just keep on sweating as I have to set everything up out there.

Across the room, Miko's already there behind the bar, but he, like Dan, also already at the bar on his laptop, doesn't look up, but when I get close to Michele she drops her polishing cloth and turns to give me what I have come to lovingly know as The Full Michele. Like Flavio, one of my best friends from college, Michele is a world-class competitive hugger. There's no negotiating with Michele. Her mom died recently and she's seen plenty of dark days otherwise, and she really and truly doesn't give a fuck about what you think you do or do not want—she *knows* you need a hug and a hug from her and she will God damned well give one to you.

We hired her not long after Sofia left The Brewer's Table to manage first a sketchy pasta bar in Dinkytown near the University of Minnesota and then, when after one day that didn't work out, one of the branches of a chain of sausage-and-beer restaurants that would end up arriving upon and departing from the MPLS/STP food scene in almost the same instant. Everybody kind of expected her to come back at any minute, but so far she hasn't.

"Hey, sweetheart," Michele says tonight, her arms reaching all the way around to my opposite shoulder blades.

"Hi, sweet pea," I say, un-aging a year or two, and then we get back to work.

I drop my bag by the bar, say hey to Dan and Miko, who reply in turn monosyllabically, and then head by Cash to face the music on the patio. As I pass, Cash pauses and gives me an earnest handshake. However rough my day and night before has been, Cash often seems to have had a worse one, but that never colors his enthusiasm for the shift ahead and the privilege that comes with it; namely, that we get to work together in such a kick-ass restaurant.

Cash, like me and, surprisingly, nobody else that I know of at The Brewer's Table, is a writer. Only, Cash, unlike me, these days writes "adult literature." He draws inspiration from Cormac McCarthy and William Burroughs and other great writers, but he is also savvy about the marketplace demands, and he's seen on Amazon where, from seemingly out of nowhere, writers of brief pornographies would spring up and make a splash and quite a bit of bank.

"How goes the ditchdigging?" I ask as we shake.

"Dude," Cash says, his big, brown eyes as eager as a retriever's, "gotta tell you about this latest one. There's this like noir angle, but it's the opposite of what you'd expect because from the research I've done the readers are still predominantly women, so instead of a guy detective and a dame client, I've got this lady detective, and she's got this guy who comes in her office, looking for someone to help him find his—his . . . McGuffin. I don't know. And that's where I am right now. I can't figure out what he'd be missing that wouldn't be super literal—like his dick or his wife—that a sexy lady detective would also be willing and/or eager to help him find, and then still for them to get busy in interesting ways, which is obviously the point. Anyway."

"Sounds great, man," I say and shrug toward the patio and tell him we'll talk again soon.

About my own writing, I don't say much, because, of course, I haven't published anything in quite some time and, after all, I'm not here to be a writer. This is not "pith-helmeted" research, as David Foster Wallace put it, that I'm doing here. I'm here to be a server because I still need the money.

That, and I have gotten used to not being as broke as a hammer without a head.

Since working at Surly for the last year, I had relatively quickly made up for our shortfall from my sabbatical and had actually begun to pay off our credit cards. The last time we were all able to go on a trip together was when we had like twenty-seven dollars to spend over a three-day weekend in LA. This year there's a conference in Reykjavik, and not only are we all going to be able to go, but we'll get to spend another five days touring (by camping and eating out at relatively nice gas stations) the coast of the whole country.

For that and a whole raft of other reasons, I couldn't and didn't want to give up serving at The Brewer's Table.

So I hit the handicap button that automatically opens the patio door, and I begin to wheel out the big black-cloth-shrouded cart loaded with all of the plates and place settings and glasses to set up the patio. The sun is brutal and unmitigated, but it is what it is, and given that patio season in Minnesota is as brief and unpredictable as a bout of hiccups, everybody takes what they can get when they can get it.

Since it's still a fairly industrial part of town, the tables get dusty just from the ambient wind between shifts, and it takes two buckets of sanitizer and two clean towels to do the job right. Each table needs to be wiped down with the first bucket/towel and then again with the second batch. Since it's so hot, the tables dry almost immediately, and I can then start slinging the bread-and-butter plates to each place setting and then place the pre-folded linen napkins with the fork tucked in on the left and the knife on the right, making sure all the while that there aren't any smudges or fingerprints on any of the silverware.

If it were any cooler than the ninety degrees it is today, ordinarily one of the other servers would have sucked it up and come out to help me stage the patio, but since it's so bruising out, everybody inside—all of whom I can see clearly through the huge windows—keeps themselves busy polishing the four or so tables in their section that were already set and more or less ready to go when they were turned last night—everybody except for Esme who now joins me on the patio.

"Esme," I say, "don't worry about it—I'm almost done."

"I'm not worried about it," she says, grabbing another stack of napkins to finish off the tables I haven't gotten to yet.

She's a funny one, this Esme. Not only is she super meticulous when it comes to studying up on the dishes and the beers, but she also has, like my sweet wife, a really rigid moral code. If there's grunt work to be done and you've got hands to help pitch in, Esme is going to be there to help with it, regardless of what it is. Usually a server's duties are limited and articulated by the manager, and there they begin and there they end. Once you've done what the manager asked, you can loaf a little before service and drink coffee or, if you're worried about getting assigned more duties, busy yourself with continuing to do the original tasks or, otherwise, just walk around briskly as though you're actually doing something urgent.

Not Esme. Like today, when she could see there were more tables to set outside and discovered that there were already five other servers folding linens in the boardroom, outside she was going to go.

After she helped me out, Esme noticed that EZ and Tiny Dancer (his regular name is Dan, but given that Dan's already our manager's name and there's another Dan in the kitchen and the fact that he's quite short, Tiny Dancer it was—I don't think there's any greater Elton John connection), anyway, EZ and Tiny Dancer are all now helping Maggie hull fava beans, and so even though she's not a cook she dons an apron and goes behind the line to pitch in.

Whereas I or other servers bring candy on Saturdays, Levi has begun to bring some obscure specialty coffee from far-flung corners of the world and brews a pot or two for us.

Before I chug a glass of water to cool down from my patio prep, I ask what's percolating today.

"Hey, yeah," he says, "so. In West Java, Indonesia, there's this coffee plant that has attracted the attention of one of the local cats, the Palm Civet, and what happens is that the Palm Civets, when the coffee trees bloom, they eat the blossoms, and ordinarily you'd think that was that but there's something miraculous about the Palm Civet's digestive tract, and the enzymes or whatnot breakdown the coffee cherries and the pulp but they don't digest the coffee beans, and so, you know, after a little while longer they, well, poop the beans out. And then the coffee traders harvest the bean poop and separate the beans from the, well, poop, and then we have the rarest, most expensive, most sought-after coffee in the world."

"You have got to be kidding," I say. I've never heard of anything quite this ridiculous, but given all of the insane things we've been serving at The Brewer's Table, I'm not really that surprised. Levi is, after all, about to load the coffee into the conical burr grinder that measures things out by grams so that he can then put it into our tedious but lovely Chemex pot. "That's what you brought us?" I say. "Poop berry coffee?"

"Oh, hey, yeah, no," Levi says. "I'm a visual artist! I can't afford that! No. This is just a really nice Indonesian Sumatra dark roast where they wet-process, or "giling basah," the coffee. I got it on Amazon."

"Sweet," I say, and thank Levi and head into the boardroom to change.

Even though it's a mixed-gender crowd, I don't really feel all that shy about changing in front of everybody. After working together for more than a year, we've come to regard one another much in the same way theater people regard themselves, which is to say, in this line of business, you're bound to see it all, and, accordingly, so long as we've got a show to put on, it's all in the line of duty. If somebody's got a dryer sheet sticking out of their jeans or an errant bra strap showing or a collar turned up wonky or a piece of spinach in their teeth, we're all obligated to say something about it.

Also I'm a very quick change since all I need to do is jump out of my shorts and into my jeans and then out of my T-shirt and into my dress shirt.

"Nice boxers, dude," Cash says. "What're those, little robber squirrels?"

In my ordinary life—teaching a class or coaching my kid's hockey team—I would've been mortified at somebody pointing out my underwear. Just recently during one of my son's games I had to get down on the ice to fix his goalie skate, and Ted, one of my fellow coaches, said after, "You put on quite the show down there. I almost dropped a coin in the slot." I was so ashamed. I don't know why it's so different here. It just is. Probably has something to do with the amount of trauma you go through on most any given shift. No matter how much poise any of us pretend to have, we've all seen each other deep in the weeds, as industry parlance goes, and it's never pretty.

"Thank you very much," I snap back at Cash. "Anyone can plainly see that they're raccoons and that they're burglars, not common robbers."

"My bad," Cash says.

I grab a seat next to Claire and start folding napkins. "So, Claire," I say, during a lull in the conversation. "I had a dream about Jorge last night."

"Oh no," Claire says. She's just returned to us after teaching for the school year in Lyon. Of us all, she seems by far the most worldly—but she's also now married to her goofy bike messenger they all call Playgirl.

"Yeah," I continue, "so he had gone on another one of his benders and—"

"Uh huh."

"Yeah, so he had gone on another bender with Dustin after some big cooking demo or something and after too much mezcal and weed—"

"Oh no," Claire says. Everybody in the room is smiling, having a hunch of what's coming, but since Claire has known Jorge the longest she also has the most knowledge of what he was capable of. Just last spring he had gone on vacation to Cozumel or someplace and had willingly posed for—and posted on social media for the whole world to see—a picture of himself wearing nothing but a Speedo that said Surly right where, well, you'd expect/fear it. (And I don't mean the butt.)

"Anyway," I say, "before either of them knew it, they were at work again as though nothing had happened, and as I was prepping the counter for service, Jorge bent over to pick up his pencil, and right there above his underwear, peeking out from the bottom of his chef's coat, was a brand new tattoo in that Olde English font, the ink still puffy and wet, that said 'I DON'T TAKE NOBODY'S SHIT!'"

Claire's poise shatters, and she begins to laugh uncontrollably to the point of crying and almost dry heaving.

There is no day so good as to be unimprovable by Claire's laughter.

Everybody else at first just kind of snickers, but Claire is just one of those compulsive laughers who, once she starts, can't stop, and soon even people who weren't in on the whole story start laughing.

Cate starts laughing in her crazy, falsetto tee-hee-hee way, intermittent with these little gasps that sound like a chipmunk hiccupping.

Madison, our new wait assist from Montana, starts laughing like nobody I've ever heard. She actually laughs "ha ha ha ha ha"—like she's saying the words and not actually laughing. People at first think she's mocking whoever told the joke—it's very disconcerting—but there's no one more kind-hearted or genuine than Madison.

Dan peeks his head in the boardroom door. "What the fuck happened in here?"

Cash, the only one of us who isn't laughing uncontrollably, just keeps folding and calmly nods at Dan. "Hey, boss. Just laughing at Matty's underwear. They've got criminal rodents all over them. Probably ought to send him home to change."

Dan doesn't react in his game-face kind of way. "Pre-shift in five," he says and shuts the boardroom door behind him. But then he peeks back in and smirks.

By the time Claire pulls herself together enough to head over to the counter for our pre-shift, her tears have subsided and her breathing is less ragged, but she's still red-faced as though she got tricked into doing shots of hot sauce.

Sensing that we're all convened, Jorge turns around from the notes he was overlooking, scans the staff, and stops. "Jesus, Claire. What the fuck is up with you?"

"Mmmm," she begins, almost weeping, "it's just allergies. I took a Benadryl. I'm . . . I'll be fine."

"Fucking better be," he says, giving his head a shake and putting his pencil decisively behind his ear. "Busy night ahead of us. We've got almost a hundred on the books and there's a show at the TCF Bank Stadium, it's nice out, and they're already running a three-hour wait downstairs." Jorge looks around again, gives the still-struggling Claire another head shake but then kind of laughs involuntarily and makes sure his pencil is still there and then keeps rolling.

"Doors open in ten minutes and we've got a lot to get through so listen up. The octopus is the same as it has been this week. Dustin's finishing one up for—who? Madison and Matt and anybody else who hasn't worked yet this week—and you got the specs on those in your email so you ought to have that nailed. Anybody?"

I don't correct him that I actually worked the night before so I had already seen the dish. It doesn't really matter anyway. In restaurant-land, there is only the night ahead of you.

Esme's hand shoots up, but everybody knows that if Esme has to do the specs on a dish, it means that nobody has them nailed yet. It's been a while since I've been called on, and I ordinarily only like to try out the specs in front of everybody after I've been able to serve the dish a few times, but I can feel the pressure since Jorge has already mentioned my and Madison's names, but then, just as Dustin pushes the dish through the pass, Madison speaks up.

"Okay, so, this is our Spanish cephalopod, and it's compressed and marinated in a salsa mancha—"

"No *n*," Jorge says, but not irritated, just correcting. "Macha. Salsa macha."

Madison keeps rolling, naming off at least a dozen of the fifteen or so ingredients in this sauce and the dish, including peanuts, guajillo chiles, pasilla chiles, chiles de arbol, garlic, and so on, but then, after remembering the mushroom and chickpea puree on the plate as well as the dehydrated chickpea crumble, she fumbles, as we all do (except

Esme and Charles), with the finishing touches that usually look like a little drop of gel or a scoop of foam or, as in the case here, a little dusting of something.

"All right," Jorge says, not wanting to prolong Madison's suffering, stern but still not blaming. "Anybody got the dust?" He looks past Esme and Charles and Dan and thinks about calling on Claire but gives her a bye. "Matt?" he says, even though he hadn't been looking at me. "What's the dust?"

I can hear my heartbeat whoosh in my ears, and even the little raccoons on my boxers squirm. The pressure of Jorge and Dustin and Dan and everybody staring at me, waiting for me to answer, feels like at least as much pressure as I've ever felt. It reminds me of being put on the spot in other restaurant jobs, but outside of The Brewer's Table, all the other questions were so much more pedestrian—Do you have wine? Are there anchovies in this? Is there gluten in your bread?—but the closest I can think of to feeling the same amount of pressure was when meeting my future father-in-law for the first time, interrupting him as he worked on a car on his farm with his head under the hood, or even during my PhD exams when one of my examiners who, it turns out, was hypoglycemic and hadn't eaten all day demanded I talk about the role of the World's Fair in Benjamin Franklin's autobiography and then later the significance of tears in Dante's *Inferno*. I did well enough with the Franklin question but then fumbled and said that Dante wept when Virgil told him of the fates of the young lovers.

"No! No! No!" my inquisitor yelled, actually pounding his fists on the table. "Virgil specifically forbade Dante tears. 'There is no place for pity here' the Mantuan poet extols."

I passed my exam, but I aged many a year waiting outside the room while the panel deliberated for the better part of a half an hour.

Here, too, can I feel myself shrivel right in front of my peers and employers, and though I had studied, about 90 percent of the ingredients for this week are all new to me. Still, I have to take a chance.

"Sumac?" I try.

"Right," Jorge says, neither pleased nor unpleased. "And what about beer pairings?"

This one would've flayed me a few months earlier, but I've been studying (not just code for drinking, thank you very much) and I am ready.

"Well," I start, "my go-to, kind of like with the tamale, would be the Fiery Hell since the puya chile in the lager brings some heat to the dish, kind of like a dash of Cholula, but for more of a soft touch the Witch's Tower would be cool—the herbaceous notes of mint and the cardamom aren't too matchy-matchy, but I'd like to see what they would bring to the dish—but we did that a lot with the tamale, so then how about Bender or even Coffee Bender? The cacao, orange, and chocolate notes might work."

The rest of our pre-shift flies by, and as I station myself at the end of the counter to make grids on scratch paper for service, Claire passes by and leans in and says, "Way to not take no shit from anyb—" but she's lost it again, and I'm worried that she's going to need more time than we have to pull it back together before the doors open.

It's moments like this where I am keenly, immediately aware of the fact that my life—like, I am willing to wager, many of ours—contains way too few moments of pure, unadulterated joy. This job, this place, these people—we are all here to make sure everybody has the best time possible—even, maybe especially, ourselves. And that is a beautiful thing.

Tonight we've got a new Katie and a new Hannah hosting, and while I like them both almost as much as the previous Katie and Hannah (I'm not mocking them or making things up; we literally replaced our old hosts with new ones bearing the same names, intentionally or not—who knows?), but they manage to seat three out of my four patio tables all at once before they've seated any of Esme's three.

But that's fine, that's okay, I tell myself. Treat your section like one big table—they'll see you do it, they'll understand you're trying to maximize your efficiency, and it'll all be cool.

My first two tables—both couples who are clearly just happy to be above the din and clamor of the beer garden below, the rough equivalent of a battle between the Danes and the Geats with a Grendel or two thrown in to help stir the pot—they both give me their beer orders and look laid back and ready to just spend plenty of time to just be outside. The third table, well, not so much.

It's a party of six, two of whom are kids. At a table that ideally only seats four—your basic rectangle with two people on each long side—they've also got one person at each end. Which would be great if they were in their own presumably capacious dining room, but the patio is only so deep, and given that we've got seven tables out there, it's also only so wide. So, God bless Katie and Hannah right down to their white cotton socks, but our new hosts have allowed this six-top to sit where only a four-top should be. And, for added fun, they've done so at the first table on the patio so that every subsequent table now involves needing to shimmy past them to get to the rest of the tables out here.

And, as though things weren't fun and tight and crowded and now busy enough, it appears that our diners are not tremendously experienced speakers of English, and though I have greeted my other two tables first and already have their beer orders, I find myself stuck at this third table struggling to describe just about everything.

As is common in fine dining across America and probably the world right now, our menus are allusive rather than exhaustive. They're meant to tease and entice rather than regale the diners with wide swaths of exposition and description. Rather than have a hundred words to let diners know what they're getting into, we have something like seven or eight.

So, instead of the long description of what's actually in the dish, our menu says simply, "Octopus, salsa macha, garbanzo" and then a number to indicate the price.

But to the uninitiated, I suppose this number could also be the quantity of octopi or the spiciness or the size party it would feed.

It takes us all quite a while to cobble together an order for them, and I'm pretty sure they don't have any confidence in me or what they ordered.

It feels wrong and presumptuous to assume that they just wanted burgers and fries and some Cokes and beers, but that was my strong suspicion, and given how long the wait was downstairs, there was no way they were getting that today.

I squeeze back inside where, it seems, the entire staff was enjoying watching me founder with my nearly full section, as they're not even faking being busy since there isn't a single diner inside the restaurant.

"What do you need, sugar?" Michele asks, knowing, despite her smile, that it isn't a fun way to start this likely nine- or ten-hour shift.

I ring in drinks for my first two tables and then think I'd better talk to Jorge before I order anything for the six-top. "When they come up," I tell Michele, "please drop my drinks at 12 and 13 and see if they need anything else. Thanks, sweet pea."

I pull up to Jorge fully prepared to unleash the ungenerous version of my six-top, but as I walk up the line, I remember not my terrible dream of his lower-back tattoo but the fact that he, or at the very least his parents—like many of the staff at The Brewer's Table and restaurants everywhere—is a first-generation American and that he or his parents, like my six-top, have at some point found themselves in a situation where they probably didn't know how to read the literal language, never mind the more inscrutable codes they form.

"Chef," I say, "may I?"

"Yeah, yeah," he says, not looking up from Instagram.

"So, I, ah . . ." I try to be decisive but it's wriggling away from me. "Table 11—the walk-in six-top—they . . . I'm pretty sure they mean to be downstairs but that the hosts down there convinced them to come up because of the no-wait situation and, anyway, I'm doing my damnedest but I just don't think they understand what our menu is. They're new to the language and whatnot."

He looks up and turns his eyes but not his face yet. His eyebrows go up and he puts his phone down, and then he finally turns to regard the table on the other side of the window.

"What's your take?" he says, noncommittal.

I sigh and take a stab. "Maybe let's order one of each small plate and a couple of the gnocchi for their kids, and then, you know, make the bill available so they can see what they get for how much money before they get in too deep . . ."

"Go with it," Jorge says, turning away from me. "Firing one tamale, one parsnip, one beet, one octo, and two gnocchi," he barks to Dustin. "And hey," he says to me after I've turned to get their drinks. "If they don't get it or aren't into it or whatever, no big deal. We'll eat the bill."

"Cool," I say, and off I go.

Within the hour, the whole place is full, inside and out. Since Lemmy from Motörhead passed away last year and Prince just this April, their music has been in heavy rotation as it is tonight, but there's so great a din that you can barely hear either of them.

Cash is the second one of us to get crushed by the hosts, and before it's even seven he's got six deuces and one pair of tables that's been smushed into a four-top.

As has become our routine, once I get full in whatever section I'm in and I've had my proper infusion of kombucha and sparkling water, I realize I can't get any busier than I am and a sense of peace comes over me and I'm able to pinch-hit in other people's sections, usually Cash's as he's got almost twice as many tables as anybody else.

We've been working together long enough that I know pretty much what Cash's thinking with every beer he orders, and Dan has it worked out through the Micros that I can tell if somebody at Cash's tables ordered a beer with a course in the pairing menu or if they just ordered one à la carte. If it's the latter, I just drop it off and confirm that it's what they ordered. If it's with the pairing menu, I give each guest the briefest of spiels as to why Cash paired which beer with what.

For reasons I don't really understand, I'm pretty much the only guy who does this for Cash. Maybe it's because the other servers don't know the difference between us. Maybe it's because he, like me, tends to find a slightly higher and more intense gear when we're slammed. Who knows? What I do know is that pretty much every time I do this, I'm able to do so on my way to running my own drinks or food or whatnot, and it basically costs me nothing. But every time he says, "Thanks, brother," it makes me feel like I really have one.

After the first couple of hours of service, the sun has finally begun to relent and I've turned my entire section outside. The couples remained happy. The six-top remained confused, but, as with Jorge's sentiment, we ended up comping their whole meal, and everybody at the table— even the kids—shook my hand on the way out.

It wasn't a success, but neither was it a loss, and now my section is full of the more typical weekend crowd, and that appears to be how the rest of the night is going to play out. But that's always what you think.

The next ripple in the evening happens when I'm inside waiting on drinks. Esme has been at the service station, but she's flustered, like she's waiting for something that won't come. She's not even able to pretend to be doing anything other than steadying her tray of drinks and staring at table 61. From what I can tell, there's nothing special about them—they're no celebrities or anything that I know of—just a couple of guys looking like they're enjoying each other's company, but then Esme approaches, drops off Frisson, one of our newest Champagne yeast–infused lagers, to each of them in a proper bubbly flute—no idea where she got those—and then the one guy drops to his knee and pulls a tiny box from his vest pocket.

"Ace of Spades" is muted from the PA and all motion stops for a second, and I swear the entire brewery can hear him say, "Will you marry me?"

And then, with just as little warning, the other man says yes, and again the entire brewery returns to party mode, Danny cueing up "1999," and I've almost never felt as happy for anybody in my life as for those two boys in this one place. Even Jorge has to wipe his eyes.

Another hour or so later and things are thinning out and it's just me and Cash and Claire and the kitchen. I've started taking tables inside because it's finally too cool to seat anybody without supplemental clothing on the patio, and I'm kind of on autopilot at this point, going through the fish details and the basic menu choices and beers and whatnot with a table that's never been here before, when I notice Cash greet an older couple a few tables away at 33. All looks well enough, but his body language is more generous and softer than usual, and I assume that means it's one of his regulars, but then the woman leaves, presumably for the bathroom, the man hails Cash back over again, and after a few words, Cash puts one hand on the table and the other on the man's back and says something I can't hear.

In a moment Cash gets himself together, nods to the man, and goes to Jorge on the line.

I'm replying to one of my least favorite questions to my new four-top—"Which of your beers is the most like Fat Tire?"—when whatever Cash is saying to Jorge strikes him, and Jorge calls to Dan, who's down at the far end of the line.

The three of them confer, and I swear I can see them all swallow something they can't quite get down.

First Jorge goes over and greets the table. He shakes the man's hand and holds his shoulder like Cash did and then kneels down and gives the woman who just gotten back the longest hug I'd ever seen. He pats the table a couple of times and returns to the line.

After I get my table's drink order and go punch it into Micros, I look around for Cash to get the lowdown, but he's nowhere to be seen. I decide to find him before I run any more of his drink orders just in case there's something I need to know, as there's one Frisson and one Diet Coke waiting to be run to 33.

I find him in the boardroom, and he's just sitting there, blinking, his arms crossed at the big table still strewn with our half-folded linens and mid-shift snacks and water bottles. I close the door behind me.

"Hey, man," I say. I hold out the little coffee cup with Sour Patch Kids toward him, but he shakes his head. "What's up?" I ask. "Everything okay?"

"Table 33," he starts. "They've been in a bunch of times. Mostly I've waited on them, mostly at 33—they like it right by the window there but still far enough away from the kitchen. I've waited on them other places too. At Bar la Grassa. Piccolo. Everywhere. I don't know how many times. But tonight," he says, "tonight is their last night—not just here, anywhere. She's fucking terminal—this is their last meal, man. They're literally going from here to hospice. She can't even eat, man. She just wanted them to have one last . . . they're just . . . I just . . . fuck."

I hang out with Cash but don't say much for a few more minutes until Dan comes in, and I go to drop drinks at my table and get their own order going.

I try to keep a soft focus on what's going on at 33 to give them the sanctity they deserve tonight, but when I finish breaking down the patio tables outside, as I roll the cart back, I find that they've already gone.

* * *

The kitchen, waxed from a hard week and a long service, has managed to scare up a few pounds of mini corn dogs and tater tots from Lord knows where—they don't serve them downstairs—and after they've had their share, they leave the rest up on the line in a silver mixing bowl as big as a kiddie pool for us FOH folks while they scour the kitchen back to a full polish. Even though they've already been dipping into the big clear-plastic Cambro filled with ice and Mickey's wide mouths, Esme takes their shifties on a tray aloft and glowing with beer like the offerings they deserve. EZ cranks up the kitchen boombox to first play what I think is a hardcore hip-hop song apt of this time of evening with the subtle refrain "Get the fuck out." From there, he puts "I'm the Scatman" on repeat at an unforgiveable volume, and before long, all the side work is done and our own shifties have been poured and, for the first time in nine and a half hours, I sit down.

Another Saturday is in the books.

Chapter Fifteen

SALT

ONE DAY A FEW MONTHS LATER, I CLOCKED IN DOWNSTAIRS and found a giant, table-sized sheet cake in the employee break room. WAY TO NOT FUCK THINGS UP! it read in cheery cursive piping on top of red and black icing. It was, I remembered, our one-year anniversary. It was the best cake I'd ever seen in my life.

As we had gotten older and our service continued to get dialed in, the food evolved into better and more inventive cuisine even though it was already as great as any of us had ever had. Jorge and Dan wouldn't let anybody rest though, and we continued to push what we were capable of by never repeating any item on the menu and by reaching further and deeper to come up with more interesting and flavorful ways to pair beer and food. The food had been insane from day one, but now we were cooking—sometimes literally—with even more special and rare beers, including a few sours. One was a wild yeast infusion called Pentagram that tasted like a fine red wine crossbred with Sweet Tarts. Another was a white wine–esque sour called Misanthrope aged on chardonnay barrels. And a third was what they call a kettle sour named Gose, brewed with sea salt and coriander. We even began infusing our beers with special herbs or fruits or vegetables or even honeycombs and pieces of wood in order to bring out or marry with flavors in specific dishes and nothing else—something we're not sure any other restaurant in the country, maybe in the world, had ever done. We had beers you couldn't have or even buy—not even in our own restaurant—unless you ordered a specific dish.

One day I was prepping my section—the counter and the banquette—measuring the distance every piece of silverware and glass was from the lip of the table so you could look down the full length of the counter and see that all twelve seats and their place settings looked exactly the same.

"It's so satisfying, right?" Dan says, gliding up to me and sighting down the counter like a sharpshooter. "To have such a tangible outlet for our OCD."

"Totally," I say. And he's right. I can think of no other place in my life where I can say that everything—literally—everything is perfect.

Of course it all changes the moment guests arrive, but still.

While Dan and I stand there, taking it all in, something strikes me. One of my duties at the Columbus Fish Market had been, at the end of every shift, to check all the salt and pepper shakers and fill any up that had been noticeably depleted. We'd been open for over a year, I realize, and I'd never done that once here. It was definitely the kind of thing anybody who's worked at a halfway decent restaurant would've known about. Certainly Dan knew about it. Why hadn't we ever done that?

I bend over and check the levels of the salt shakers on the counter, and they're all completely full. I run around the dining room and they all are too.

"What?" Dan says, seeing that I'm onto something.

"It's the salt shakers, Danny," I say. "They're all full."

"Yeah," he says. "So?"

I know Dan's OCD by now as well as my own and am completely sure he's of the same mind. It would've been unforgivable to have an unclean or less than full shaker of salt on any of our tables. It's not that he hasn't noticed or overlooked a job to give us. It's that we've never needed to have done it in the first place.

"They're full, Dan," I say, smiling broadly as though I was just about to invent the cure for cancer. "They're all full. And when was the last time we had to top them off?"

"Jesus," Dan says, understanding. "Never."

"We have literally never had to fill the salt shakers because—"

"Because nobody ever uses salt here."

Without even asking his permission, I grab a tray and go gather all of the salt shakers from our tables, still beaming with the realization that this is how perfect our food is. Not only do we not need to check the levels of the salt shakers. We don't even need to have them on the table.

Some restaurant's food is so blandly prepared or its flavors so critically muddled that every table has those actual buckets of salt and pepper and hot sauce and barbecue sauce and ketchup and mayo and mustard and so forth on them. On our tables from now on, all we'll have at each setting are a fork and a knife tucked neatly into a perfectly folded linen, a spotless water glass, and, for the table, a glass hurricane vase partially filled with barley, a candle tucked on top of it.

I can barely express how happy this makes me.

Of course, there were growing pains too. I got yelled at by Dustin for assuming once that the kitchen wouldn't want to split a dish for a couple that wanted to share something, and we collectively had one bad shift that winter where at least three or four steak tartares were made instead of the sturgeons the guests had ordered. Jorge was so livid about this he demanded we meet immediately after service, but there was nothing really to say. Hans had recently quit because he was convinced he'd make more tips on his own, and so we had Mikey, a pinch-hitter server from downstairs, working that night, and we were all pretty sure it was his error, but none of us were willing to blame him, and I think Jorge could sense that, so in the end he just shook his head and smiled. "I guess the moral of the story is," he said, "don't write down 'ST' in your books for the steak tartare and the sturgeon. I don't know . . . write 'STU' or some shit. Just don't mix them up again."

We'd also seen some turnover by then—as is common in all restaurants. But some were more upsetting than others. Christina we all missed dearly. Hans had been pretty aloof from the beginning, so his departure wasn't really a surprise. Rikki left to go work at one of the many new breweries by Target Field, which was a loss for us but felt like a good thing for her—our menu and all of its particularities had really begun to stress her out. But then, in a move nobody knew much about or saw coming, Linda was suddenly gone and Dan had been promoted to replace her as the head of hospitality for all of Surly dining. He was still in the building, but he wasn't just ours anymore.

And then, a bit more personally, as our second year came around, Miko shocked us too when he decided to take a job at a start-up sour-only brewery in Chicago, and then Claire left to teach in France again, and Cate headed out west to take care of a bunch of her wayward friends in San Francisco—we all cried hard on her last day and almost everybody got goofy friend tattoos with the words "Plot Twist!," a favorite saying she and Miko would holler upon just about any new development in the evening. But despite all of these shifts and transitions, we remained who we were at the core.

We still had Jorge and Dustin. We still, more or less, had Dan. And we still had Cash and Levi and Esme and our new beloved Michele and, well, me.

I had been teaching again, but since I was only scheduled on Wednesdays and Saturdays—and had quit the lake place at the end of the summer—staying on was not only something I felt like I could do but also something I had to do. If push had come to shove, I had thought, I would quit teaching before I'd leave The Brewer's Table.

Our new hires were all brilliant and super talented people, and I was stoked about our new manager, Emily, who was from my hometown of Milwaukee. As soon as she started, she brought a new array of plants to decorate the restaurant that gave the whole room a kind of terrarium vibe, and we finally got approval to put a couple of Levi's pieces of cool industrial artwork on the walls. On top of it all, somehow I had become in charge of managing the tip pool arrangements as well as training new employees, including Emily. I had never trained anybody at anything before and found that I absolutely loved it.

One April day, just a month away from our second anniversary, Dan sent everybody a message that there was a mandatory all-staff meeting the next day—a Tuesday morning. We'd been having all-staff meetings for a while. They included the beer hall servers and The Brewer's Table staff, but as most of these meetings were steered toward basic food and beer knowledge and I had begun to skip them, pretending that I was teaching at that time. Nobody ever said anything about my absences, and I expected this one would be the same, but right when the meeting was to begin, Dan sent me a text.

"Where are you? Meeting's starting. Get here ASAP."

I think about it for a minute. If I don't reply it can mean anything. If I do reply it means that I'm choosing to be somewhere else over being there. I think about it for another minute, staring at my phone like I have to unlock a codex to get it to work instead of just pressing a button.

"Are you coming?" he texts again, making me nearly drop my phone. "Everybody's here. This can't wait."

Shit, I think. Something heavy is going down. I should hustle there, but unless it's a long meeting, even if I leave immediately, still wearing my PJs, I'll miss whatever was so urgent. I decide to pretend I didn't have my phone on and go for a run, but as I pad around the lake, guessing at what the meeting was for is all I can think about.

About an hour later I can't take it any longer. I need to know if more people are leaving or getting shitcanned or if there was more disciplinary harshness from Jorge over more ordering snafus I hadn't known about, or maybe I had fucked up the till, as I had recently started bartending after Miko left. I had thought everything was okay, but because there are so many moving parts to a proper restaurant, anything could go wrong at any time.

"Hey Danny," I text. "Sorry I couldn't make the meeting. Had one at school that didn't let out in time," I lied.

"No prob," from Dan and then nothing else for what seems like a small eternity. I'm standing in my living room, Frankie, my Aussie, staring at me as I stare at my unresponsive phone.

It feels like a weak move, but I have to write back.

"Everything okay?" I ask.

Nothing again for many minutes. Frankie whimpers, urging me to sit down, but I can't. Dan always has his phone on him so I know he's just fucking with me by not replying, but there's nothing I can do. Finally those three little dots appear in advance of his text, which means he's typing.

"Yeppers."

Yeppers. What the fuck kind of a reply is that? I mean, it's better than the most dreaded text message reply of all ("k" or, arguably worse, "k."), but it was still pretty impenetrable.

"That's a relief," I write. "Was worried. So it's not bad news?"

"Nope."

Oh my fucking God. He's killing me, I thought.

"So it's good?" I write.

"Yep," he says. "You could say that."

FOR FUCK'S SAKE DAN! I type and delete three times, during the last of which he texts again.

"Just a cool review," he writes. "See you tomorrow."

Like I should've done an hour ago, I Google "Brewer's Table," and the first thing that comes up after our own website is confusing. It's not anything from local media, which has by now pretty thoroughly and warmly reviewed us. This latest bit is from *Food & Wine* magazine, and the headline is simple but it reads to me like something in a foreign language: "Food & Wine Best Restaurants of 2016. Start slideshow."

Food & Wine? I think. What the hell could they have to say about us? Has *Beer Advocate* suddenly agreed to start rating destination wineries if *Food & Wine* reviews breweries? Not a chance. So I start the slideshow. And, sure enough, there we are. It's us. Us and nine other restaurants from LA to NYC to Washington, DC, to Chicago—there we fucking are. Among the top ten best restaurants of the year in *Food & Wine* of all places.

"Holy fucking shit, Danny!" I write.

"Yeah, pretty cool, eh?"

"That's one way to put it."

I swear I can see his beautiful wry grin sparkling like the results shot in a teeth whitening commercial all the way through my phone.

The next few weeks were mayhem. We were busier than ever—booked solid for a couple of months in advance on the weekends (though those Wednesday and Thursday nights were still a little sluggish). And initially it was great. Everybody was so stoked for us and just happy to be able to snag a table, whether right at five or at our latest ten o'clock seating. Things mostly felt familiar, just busier, but at the same time we also had a lot of newcomers, many of whom made it abundantly clear that they were skeptical about coming to a brewery for a meal. And though many people now anticipated an unparalleled and unforgettable dining experience, a few took the rave reviews as a kind of dare or taunt, and I'd hear rumblings as I passed by tables—nobody ever said this kind of thing directly to anyone in Minnesota, but once I overheard someone say, "That's hardly what Escoffier would call a panna cotta" and, another time, less adroitly, "Best restaurant of the year my ass."

We were on top of the world, except we kind of weren't.

Rumors started going around that Linda hadn't just left. She'd been fired. Not long after, Todd announced in a terse press release that he was moving to Chicago for a collaborative endeavor with another brewery, Three Floyds.

From what I could piece together between rumors and hearsay, despite having been in business together with Omar for ten years at what by now was the biggest destination brewery within a five-hundred-mile radius, Todd was just his employee, not a partner. He still clocked in, just like I did (but he was number 1, whereas I was 500). We were not, it was painful to realize, an employee-owned brewery like Odell, Deschutes, Oskar Blues, or New Belgium. Surly, apparently, was a kingdom with no hierarchy. Just a factory that made beer plus a throne for one. Despite everything Todd had done for Surly—from starting things up with Omar from day one, brewing that pony keg of Bender in Omar's driveway on New Year's Eve, to coming up with Furious and Darkness and all its other best-selling, award-winning beers—in spite of it all, Todd, I've heard, had never received anything more than a paycheck for all of his innovation, hard work, and dedication. From what was said around the restaurant, he wasn't even allowed to be a partner or own any stock in the company. Soon after he left, his namesake legacy beer, Todd the Axe Man, was stripped of his name down to simply the Axe Man. It sounded like as bad a betrayal as Tesla by Edison. Todd was clearly right up there in terms of innovation with Nikola Tesla, but Omar, unlike Thomas Edison, apparently didn't really have much to do with brewing and now drove a hundred-thousand-dollar Tesla. I know it's a jumble of a metaphor, but all I knew for sure about either pair of partners was that it was a damned shame one guy was getting most of the credit—and money—and the other guy had to hit the bricks.

But to hell with it, I thought. We were bigger and better and more ambitious and artistic than any personnel changes, however small or large. And we had other bigger and better news hopefully coming soon, as Jorge had been nominated for a James Beard award the year before as one of the best chefs in the Midwest and this year he was up for it again and was in the finalist round. Only five chefs win each year in the entire country—and we all had at least a couple fingers on each hand crossed for this achievement. The *Food & Wine* thing was supercool, but it was basically the fine dining equivalent of Guy Fieri deciding to feature your place on *Diners, Drive-Ins, and Dives* (which he had done for the beer hall a few weeks prior). It was great publicity, but it wasn't anything you really were striving for. The James Beard award, though, that was epic. That was the chef's Pulitzer. You were basically golden for a long while after you won that.

With the Beard announcement still pending, on the day before the end of May, Dan—not our new manager, Emily—called another unscheduled all-staff meeting for The Brewer's Table. This one I knew I couldn't miss since I had lost out on all the champagne they all shared at the *Food & Wine* announcement, so I got there early and found the dining room full of our team, lounging on the banquette and the four-tops, chitchatting and trying to pretend like we weren't all watching Jorge and Dan for any signs or tells as to what the big news was.

After the *Food & Wine* news, short of a collaboration with a multiple-Michelin-star chef like Daniel Boulud or René Redzepi, I couldn't imagine what else could be worthy of an all-staff cattle call. Since we didn't live in a city where there were Michelin Guide books, so long as Jorge lived and worked here, he'd never receive a Michelin star, and so our *Food & Wine* rating and, hopefully, this James Beard award was the best it was ever going to get in Minnesota.

Jorge begins to clear his throat—his sign for us to quiet down so he won't have to shout—but before he starts I notice Omar coming through the front door. Who could blame him for wanting to be here for the big announcement, I think.

I was a little bitter about the whole Todd and Linda thing, but by this time I had served Omar and his entire family several times. And though he was usually really quiet and reserved—he'd typically arrive late and then leave early—his wife was sweet but fierce and refused to be bullied by our weird, sometimes strange food. She called me over once and gestured subtly to her plate. "Is pork belly supposed to be *this* fatty?" I didn't know what to say. I mean, it's not called "pork abs," after all. But still, what was on her plate looked like one tumid baton of lard. I tried to apologize and reached for her plate, but her look stopped me before her hand had to. She knew everybody would be watching. She just needed to know, and my gesture, I had hoped, told her she was in the right. Sometimes even fat can be too fatty.

I had waited on Omar's kids, who quietly endured our tasting menu as though they were being taught a lesson for some kind of culinary transgression. I had even waited on Omar's sweet and sassy German mom, Dorit, who was as befuddled and simultaneously amused by the weird stuff we were serving as though we had gone as far as Wylie Dufresne or that over-the-top molecular guy, Marcel, on *Top Chef*.

Sure, Omar drove a car that was about as expensive as my house and was never really vocal about his enthusiasm for what we were doing, but when it came down to it, if it weren't for him, I wouldn't be here. None of us would. About that, there was no doubt.

"So," Jorge starts, sitting on a chair, leaning forward on his knees, not really looking at anybody. Dan is on his right. Omar is on his left. "So," Jorge starts again, "we're closing."

After a few audible gasps, the only sound is the kitchen ventilation system audibly sucking any remaining air entirely out of the room.

Time blurs and I can't help but expect that this is the prelude to a prank—that Jorge really won the Beard award or that maybe the Michelin people have decided, because of us, to do a Michelin Guide of the Twin Cities and we will receive the first star of the region—but no.

Jorge does not elaborate but is nodding silently, staring at his folded hands. Dan does not speak and is shaking his head in disbelief.

Omar appears sanguine and also in kind of a hurry. He doesn't stand to speak and just jumps right in.

"You know, having accomplished what we set out to do here, it's time to seek out the next challenge." It sounds more like a corporate line than anything I'd ever heard at Surly. He goes on, barely, to say that he hopes some if not all of us would stay on to see what that new next challenge would be, but for now "the investors" had hired a new manager from Granite City—a nationwide brewpub chain that also owns Famous Dave's, Bakers Square, and Champps—to come in to tighten the screws and cut away the fluff and fat. As shocking as the news was that we were closing, it wasn't difficult to imagine upon even the most cursory glance that if all you care about is the bottom line, we were surely the fluffiest and the fattest cats in the building.

240

Omar gets up brusquely and leaves the meeting without any sign of emotion other than awkwardness, and he waves a dismissive thanks as though he just gave a talk about brewing to a bunch of fourth graders. Like them, many of us are sitting cross-legged and baffled on the cold hard floor. Dan and Jorge, I notice, both begin to bob their shoulders in advance of actual tears, which trips off crying from pretty much everybody, even Celia, who was sitting next to me. I had just begun training her last week as our newest—and, apparently, last—server.

Dan coughs and begins to collect himself, and Jorge, still crying, but in a tough-guy way, gives us the option of closing immediately or working through until August.

As we debate it among ourselves, Dan shakes his head and wipes his eyes. I'm sitting right in front of him, shaking pathetically in advance of a full-blown crying jag, Celia rubbing my back to try to get me to calm down.

"Fuck, Matty," he says. "It was like just yesterday when we were interviewing. I remember you said that Furious was the 'tuxedo of beers,' and I thought 'Fuck yeah, we are going to kick some ass,' and we did."

"I remember," I say, doing that laugh-cry thing where I appear to be almost throwing up and coughing at the same time.

"This was," Dan says, looking around at everybody, "this was the best job I've ever had. I never had a bad day here. I always loved coming in to work."

"Well," I say, "there was that one day after the Limp Bizkit show . . ."
He laughs and cries and nods.

"Okay," he says. "Other than that one day, this was always my favorite place to be. And it was Linkin Park."

"Oh right," I say. "How could I have mistaken those two?!"

We all laugh and cry some more, and a couple of the cooks are so upset that they leave without a word through the back, but almost all of the rest of us just start pouring ourselves beers and mill around and try to help each other make sense of it all and to decide if we're going to be done right now or stick it out through the summer.

Todd and Linda and Omar and all of that notwithstanding, I think I can safely speak for most if not all of us. Dan's right. This place wasn't just a good job. It was the best.

But, apparently, just like Walden, Brook Farm, Biosphere 2, the early years of the Obama administration, and all the other utopias, whether they were planned or happy accidents, every perfect place sooner or later has to close its doors and kick everybody back out to the real world.

We decided to stay open until August, and though we all made good money, I think we all mostly regretted the decision. Sure, almost immediately after it was made public that we were closing, every available reservation for the whole rest of the summer was snatched up, no matter which day of the week it was. If we thought the crowd following the *Food & Wine* news brought some unpleasant new guests, the gawkers and carrion pickers who came in following the public news of our closing were even worse. Nearly every table had only come because we were closing, and thrilled though they were to know that they were getting a seat at a restaurant whose days were literally numbered, they couldn't help themselves, every day, every hour, from asking all of us if we weren't still excited to be there and if we weren't still thrilled to be busy, and then, when they almost visibly could be seen figuring out what a predictable and often repeated line of inquiry they were following, they'd start asking us if we didn't have other jobs lined up (we didn't) and if we weren't looking forward to the next big thing (we weren't) and if Jorge wasn't going to bring us all on board once he landed on his feet as surely he would (we didn't have any idea what he was going to do next, and neither did he).

The last couple of months were funereal and dreadful, but, yes, we were busy and made some good money.

Still, though, throughout it all, there were beautiful, talented, wickedly smart people at every corner of the restaurant. Our new manager, Emily, was brilliant and thoughtful and more vigorously empathetic and creative than most anyone I've ever known. Our newest servers, Sadie and Celia, and our new food runner, Laura, all became some of my favorite people ever. Cash and I were stalwarts throughout it all, and it seemed like there was nobody we could count on more, no matter how dire the shift or the hangover, than each other. He had even begun packing a flask of whiskey in his bag, not for himself but for me. Well, mostly for me.

But as the end of the summer neared, I began to get stomach cramps and waves of nausea out of the blue, sometimes on my way to work, sometimes in the middle of service. I stopped drinking coffee and then kombucha and finally even Cash's whiskey, but none of those efforts did anything but made me feel worse, sleep less, and be more fatigued and woozy on my feet at work.

I never did find out what was wrong with me, but I had a feeling it came down to a growing realization as we approached our end: I would probably never see most of these people again.

And, sadly, I was right.

Like Hemingway's notion of Paris as a young man, it would seem that the enterprise of a restaurant is a moveable feast, one that you can take with you wherever else you go. And that might be true for most of my coworkers who have bounced around from one restaurant to the next, such as Sofia, Michele, Levi, Cash and others, but I doubted it would again be true for me. I was, I had to admit, too old to do this again.

Too old, yes, that's true. But it was also an excuse. A safe lie nobody at home or at my school would argue with or even ask for elaboration about.

The real reason is that I fell in love here. I fell in love not with a person or a place or even the food or beer but with the promise of both what a family of people can do and who they can become when they all believe in the same thing—and that same thing is, simply, taking care of each other.

BEAUTIFUL FRIENDS

AND SO, AFTER BABY STEPS HANDED OUT ALL OF HIS BEER, after Jorge's frijoles con puerco were gone, like all suppers, our last supper was coming to an end. Our final service had yet to begin, but we were all already ragged from the litany of lasts we'd been going through. Whether we've said it to ourselves or to each other, all day long, it's just felt interminable.

This is the last shirt I'll wear to work, I said as I picked out a vintage Western deal with embroidered flowers all over the shoulders à la Gram Parsons.

This is the last commute I'll make down the transitway.

This is the last bag of Sour Patch Kids I'll bring for everybody.

This is the last time I'll get to use the service door by the loading dock.

This is the last time I'll clock in.

This is the last time I'll get to chitchat in my broken Spanish with the pastry cooks while I wait for the elevator.

This is the last high-five Eric will ever give me, calling me Mark as he always does as he rushes back to the patio bar.

This is the last time I'll tap twice on the swinging door into the kitchen before saying hi to Maggie as she preps in the back for service.

This is the last time I'll shout "corner!" like Pat taught me before rounding the wall dividing the front and back of house, as I head into the dining room for our last day at The Brewer's Table.

I had already broken down in the shower, and I also cried while riding my bike to work—that was a first—and even though I begin to tear up again as I stride across the floor, I know it won't be the last time I cry tonight.

Cutting into the profundity of this last service, however, is a film crew taking footage of Jorge. We knew they were going to be here, but not in such force. It hasn't been unusual for journalists and/or photographers to be in the restaurant. Between them and the food and beer bloggers and vloggers and the ubiquitous raft of Instagrammers and Facebookers, we've grown accustomed to walking into the restaurant to see somebody standing on a chair so they can get a better angle of a dish or people doing it more subtly during service, but today is different. This crew has got a complicated, multi-handled camera, a huge kite-sized lighting rig, a guy with a boom mic, and even an interviewer.

Before and during family meal, we've tried to ignore them and let whatever happens happen and also let whomever the crew has pinned in front of the lens suffer alone as we mock them from behind the camera.

But it's weird and painful to have such a personal and bitter thing also be a media event.

Since it went public that we were closing, every day became the equivalent of our busiest Saturday ever, and all summer long we've been regularly serving close to two hundred people a night. Everybody thought that it would be nothing but our favorite diners, friends, and family, but it's actually been mostly strangers, and given that we already had our funeral scheduled, it was really hard to get all that enthusiastic when meeting new guests since we knew we'd never see them again. All summer, all strangers, and we just wanted our regulars, our heavy metal brewers and beer geeks—even Roland—back.

Even though there had been plenty of turnover, there were still many of us who had been along for the ride from the very beginning, and here we were, about to bury the realization of all of our hard work and dreams. It was kind of crazy. I asked around and it turned out that, of our entire FOH staff, I couldn't find a single person who had worked all the way from the opening to the close of another restaurant. Not one of us knew what it meant to live through the entire lifespan of a restaurant.

But, finally, here we are. An hour before our last service. We knew this night was going to have an end, and, by all accounts, it's probably going to be messy, but none of us really seem to want to do anything to make it begin. Because once service starts tonight, the end is simply the inevitable extension of the act of opening.

In the lull between family meal and the start of service, Emily has arranged for us to all do a half an hour of "chair yoga." Esme, thank God, is our yogi. She has us push some tables out of the way and we each stand in front of a chair, shoes on or off—whatever we're into. It's silly and awkward at first, but after a bit some of us are able to relax into it, though only a few cooks are participating. The rest of them are back behind the line again, finishing their mise en place for the long service ahead, and as they chop herbs or pick mint leaves from their stems they collectively make fun of us, albeit at a somewhat hushed volume because Pat is taking the yoga as seriously as everything she does, and they all know that she'd run any of them up to the hilt of her chef knife if they got caught making fun of her.

"So now," Esme says, fully in her element, "we're going to try a little tree pose, but with friends." She has us stand and raise our right knee and then prop that foot against the side of our left knee. "Now raise both your hands to the side, palms out, and lean into your neighbors for support."

Instead of it being the relatively difficult and teetering pose it would be for a bunch of yoga amateurs, the support of each other's hands on both sides makes us a firm and steady circle. Except for me, because Celia is on my left but then there's a round four-top between me and Pat.

For some reason, under Celia's dress shirt she's wearing what appears to be a one-piece swimming suit. She and I put our palms together and almost complete the circle, but Pat's arm doesn't even make it halfway across the table, and so we have to kind of do air yoga until enough people notice and begin to scooch over so that my and Pat's hands can finally bring us all together.

This would be supercool and stress relieving if not for the lady from the film crew. She's lingering just outside our circle, waiting for us to sign releases as though she's an Amway salesperson or trying to get the autographs of people she thinks might be famous.

247

For better or for worse, the last pose is over in a few minutes, and Esme has us sway or dance or stretch or whatever we want to do for the last couple of moments to what I think is some acoustic cover of an Ani DiFranco song, and I go over and give her a big hug.

"Oh Matty," she says.

"I'm so glad you could be a part of this, Esme."

"I know," she says. "I was so glad when Emily asked if I had any ideas for tonight."

"She was saying some pretty crazy shit for a while," I say. Around us the circle has begun to break up, and people are putting their Docs or clogs back on and gathering their dress shirts and pants to change for our shift. Esme is unhurried because she actually left The Brewer's Table earlier in the year to work in special events and also to become the lead teacher of Surly's monthly Beer School. Both she and Dan had become official cicerones—as arduous and almost as well respected as becoming sommeliers—and nobody has hard feelings. We all love Esme, and her genius was not forever going to be contained in the role of server.

"At one point," I say, "Emily threatened us with a mariachi band. A live DJ. Maybe a karaoke machine. I'm so glad it was you."

"Me too," she says and gives me one last squeeze.

"I don't think I've ever told you," I say, feeling moved by the occasion to be a little extra effusive. "When I talk about work at home, do you know how Jenae talks about you to keep you straight from all the servers?"

"No," Esme says, "you've never said anything."

"The beautiful one," I say. "That's what she calls you. Not like it's a description but more like a title or the definition of the word. Esme, The Beautiful One."

Her sweet brown eyes shine a little brighter, and I give her one last hug and then give her over to the others who want to thank her too.

While I can get it, I decide to go out for some fresh air and to help prep the patio tables for service when I notice Jorge off to the side by himself. He's leaning against the rail of the balcony, overlooking the beer garden with a blank expression, but as I've come to learn, he's usually feeling something deep and powerful even if he can't quite find the right words to express himself. He is the consummate chef, which I am reminded means both "chief" and "head." There's no question that this was his masterpiece and that his food is what he communicates with. If we asked him to write or paint a self-portrait, he would likely founder (except on Instagram, where he's actually super emotive), but if we asked him to cook a dish that explained who he was, he wouldn't hesitate. It might be that sublime tamale or another Yucatecan dish such as huevos motuleños, or it might have been the frijoles con puerco he just made for us all.

Now, he is as alone as I've ever seen him. He isn't on his phone or looking at anybody in particular, and, as usual, I'm inclined to give him the solitude he seems to want, but I am also, like Marlow, realizing that my time to approach him in the last of his element is an opportunity I won't get again. And he, like Kurtz—albeit with a lot less bloodshed and ivory stockpiling—is certainly aware that his rule has been subsumed by powers he ultimately never actually controlled. Ours—*his*—was a wild and strange but ultimately fragile trip, and it will today come to its end.

Tess, our latest and last host, and Baby Steps are moving the tables up against the window in a way we've never done. The tables are usually against the balcony railing, but now they're up against the glass. To me, this is as upsetting as reversing the order of the keys of a piano or insisting that I write with my left hand.

Just having earned her real estate license, Tess is riding out these last few weeks with us. She's as adroit and lithe as a professional dancer, but despite the fact that she does her job terrifically well, when the nights move on and get more and more stressful as the middle turn inevitably jams up against the last turn, she gets so tense her shoulders start to draw up nearly all the way to her ears like she's doing that dance move from Michael Jackson's "Thriller" video.

Baby Steps—another recent addition after Miko and then, more recently, Remi left—is an absolute sweetheart, as well as the leader of an alt-country band. He prefers to go by JJ, but we all call him Baby Steps because, even though he always gets the job done, he does so in an excruciatingly slow way. In addition to bartender, he's also one of our wait assists, and it's his job tonight to help out whoever is assigned to work the patio. I'm glad it's not me.

I must have looked conspicuously perplexed at the two of them because Tess begins to explain without my asking. "Emily said it's probably going to rain at some point. We've got so many covers," Tess continues, "we have to have the patio open."

I have no reply. It's confusing and disturbing, and I hate that we're doing something so different on the last night. This has never been a place to just wing it and hope things go well, and here we are, doing exactly that on our last day.

Sadie is another new server and has become one of my best friends at the restaurant. Like me, she is married and has one child and a headstrong dog, and we often commiserate about kids and pets and in-laws and the like in ways we couldn't with most of the rest of the staff because most of them aren't married, don't have kids, and all of them seem to be cat people.

"What the fuck is this?" Sadie says, standing next to me, assessing Tess and Baby Steps and the emerging patio situation. I'm glad it's not just me who is bothered by this.

I feel deeply bad for Sadie, though, because it's she and Celia on the patio tonight, and not only will they be stuck out here by themselves, but Baby Steps will be their sole wait assist.

We all know each other, our strengths, our challenges, what we love, how to make each other laugh, what to do when we're having a rough shift—all without needing to ask a single question. We can read each other's body language, the way we stride or stumble across the dining room, our facility or frustration with the point-of-sale system, our proud or crumpled posture at tables. To each other, we're all open books.

Tonight, Sadie's shoulders are slightly slumped, whereas they're usually proud, up and back, and instead of holding her head sweetly high and defiant, her rolled shoulders are forcing her head forward and down. Usually she's radiantly happy and easy to laugh and, while doing whatever task at hand—polishing a glass or punching in an order—she has a way of stopping what she's doing for just a fraction of a second and looking at you and smiling. Nothing grand and nothing with words—

but the way she does it and that little wisp of her hair falling across her eyes—she stops your heart. But now, with her shoulders rolled and even with her hair in her eyes, she just looks sad.

"Seriously," she says. "What the ever loving fuck?"

"Rain," Tess tries again, straining with Baby Steps to move the last of the tables.

It probably means nothing to our diners or our hosts or the cooks, but moving the tables is everything to us servers. Our entire interface with our tables depends on knowing where to stand. Every table in every restaurant in the world has assigned numbers to every seat. Where the server stands is usually a fixed position, and then the diner to the left is seat 1, the next one clockwise is seat 2, and so on. It's how good restaurants are able to bring you what you order without "auctioning it off" at the table ("and who had the tacos . . . and who had the fish?"). To move a table—never mind all of them—in a given section is close to a server's worst nightmare because, sure, we can still stand at the head of the table and go around clockwise from our left, but Emily won't have time to change that in Micros or to fully explain it to our food runners, so it falls entirely on Sadie and Celia to calculate the new positions to the old schema. It guarantees a shift of looking like a fumbling amateur, and that's as far away from who any of us want to be tonight.

"I'm so sorry, Sadie."

"Jesus Christ," she says, shaking her head, trying to let it go but clearly not entirely able to.

"I know," I say. "I know." I pat her shoulder in a fatherly way that she shrugs off. If I do it for one more second I know she'll smack my hand away. Sadie is sweet and also a badass. She needs no one's help or sympathy. What she wants is to be able to do her job in the same kick-ass way she always has. This is a sucky thing to have happen on any day, but on our last it's almost unforgiveable.

She takes a big breath and lets it go, and something has changed inside her, and I think I know what it is. Since it was announced that we were closing, one by one, we've each begun to realize that the worst is already going to happen. "What are they going to do," we each have said of one thing or another this summer, "fire me?"

Sadie does this funny thing with her lips, pooching them out and kind of giving the patio an air kiss.

"Whatever," she says, almost cheerful now. She blows her bangs out of her eyes and goes back inside.

But nobody here has ever said "whatever" to anything. And nobody here wants it to end like this.

Nobody here wants it to end.

I turn around, again on the verge of breaking down, and I move to go inside, but I don't want to be in there quite yet because I can see the camera crew is back at it, doing something with the chefs on the line, and all their stuff is all over the boardroom so I can't go in there and there's a two-hundred-person wedding going down in Scheid Hall in the event space, which means there's bound to be a cluster of drunk idiots in rented suits and/or lavender taffeta, and I spin in a complete circle before I see Jorge again and decide, What the hell.

I feel sheepish approaching Jorge under any circumstance, but especially now considering I'm not in uniform yet and am still wearing shorts, a T-shirt, and, super awkwardly, my work clogs. Jorge, however, doesn't exactly look at me but almost imperceptivity shifts his eyes in my direction. For him, this is the equivalent of turning his entire body and holding both of my cheeks in his hands.

"Hey," he says.

I don't say anything. I try to lean casually against the rail but don't quite pull it off. It's taken all my guts to come over to a man I have known and worked with for well over two years.

"How you doing?" he asks. An astonishingly intimate question for him, especially given my lack of reply to his opening salvo.

I am totally overwhelmed. I had words for Sadie, I guess, because we always talk. Now I'm unable to coax even a single word and am finding myself overwhelmed again with an emotion I don't want and can't control. I shake my speechless head and fight back the tears only slightly more effectively than Meg Ryan in, say, any movie she's ever been in. It takes nothing for me to recall the last time Jorge cried—when he announced that we were closing—and I still remember thinking, Oh fuck. If this guy is crying, I'm going to be one soggy cookie.

Jorge and I, despite the seemingly vast differences in our positions, have a lot in common. While he comes from Latino stock and I'm vaguely Irish/English, we both grew up in the Midwest and went to overpriced private colleges and graduate school, for which we will be paying well into our Geritol years. And despite the fact that neither of us could have more public, conspicuously type A jobs, we're both brooding, contemplative, emotional guys who are, ironically, only rarely expressive. This personality type, I'm told, falls somewhere supposedly strange on the Myers-Briggs inventory, but I could never remember exactly what the letters and numbers were. "The Architect" was what we or he, at least, is called.

Standing on the second-floor balcony of this brewery and restaurant, it isn't at all difficult to think of him as precisely that. I've met the actual architects of this place, and, of course, Omar and Todd and Linda and Dan and everybody else who was in on designing what this place was going to be, but Jorge to me has always felt like the guy who is the real architect of this place.

"I know," Jorge says, allaying my apparent speechlessness. Ordinarily he'd have sent me for the break room, disgusted with my inability to properly man my emotions and my tongue. Today he seems gentle, forgiving. A line from a James Wright poem—the one about Judas—comes to mind. "Flayed without hope, I held the man for nothing in my arms."

"It's pretty cool, you know?" Jorge says. "That we got to do this. That we got to do it at all. It's been an amazing ride."

"Yeah," I muster.

In the beer garden below, there are a few of these little grass mounds shaped like perfect little cones, and on one of them an imperious black-and-white terrier stands, defiant and proud, as though he's in charge here and is proud of himself and the job he's doing.

"I've been thinking," I say, suddenly able to talk. "It was about something Dustin said. About how we're not a 'real' restaurant. About how we play restaurant instead."

I panic for a moment, unsure whether this was something that Jorge himself might have said or maybe was something Dustin said but not with Jorge present so it might sound derogatory, and I don't know how much to explain or if I should just go on or stop.

Down below, a guy is playing cornhole with his son, and his first beanbag hits dead center. His next one lands with a thud just to the right of the hole, and the kid does this little taunting dance in a circle. The dad sends the third one flying and the bag beams the kid in the face and the dad goes running to see if he's okay.

"Anyway," I say, "I've been thinking a lot about that phrase, and just this morning it hit me that it's not so much that we 'play restaurant' like it's some kind of game but rather that we play restaurant like it's an instrument—the whole thing—the building, the furniture, the windows, the light, the shadow, the beer, the people, and, of course, the food."

I take and release a big breath after my ridiculous soliloquy, unsure if I made any sense or am being irritating or what.

"Yeah," he says. "Pretty cool."

The father in the garden tries to pacify his son and he's failing beautifully as his kid is sobbing and heaving—I can't hear him but I can see his chest struggle as his fists pinwheel toward his father's face.

"I mean," I say, elaborating, as usual, more than I need to, "it's kind of like the joke about that Velvet Underground album. It didn't sell that many copies, but everyone who bought one started their own band."

Jorge nods, but I don't think he agrees with me. He probably just wants me to go away. His short-term future self would certainly not agree.

After The Brewer's Table closes, Jorge will go without work for quite a while and then on to briefly cook at a tequila distillery in La Crosse, Wisconsin, only to then come back to Minneapolis to work on a Mexican pop-up concept called Pollo Pollo. Years later he'll open a couple of restaurants for developers in Toledo, Ohio, of all places, and then in the old Blackbird space on Nicollet, his first restaurant truly of his own: Petite León. But that will be years from now, and all that the Jorge before me knows is that this ride has come to an end.

Dustin will return to the job he had before he came to Surly. A couple of our line cooks become line cooks downstairs. Esme and another couple of servers will stay at Surly in varying capacities, but most drift off to different jobs, never to work together again.

There will be no big happy ending for most of us at The Brewer's Table. This is the end.

One of my and also Emily's favorite bars in Milwaukee is situated on the first floor of a house. It became famous, in a way, for handing out bumper stickers at the end of the night that said "I Closed Wol-ski's." You couldn't buy one, but if you were there for closing time at 2:00 a.m., you got one for free. I've encountered such stickers in Bos-ton, New York City, Ireland, even Iceland. For our last day, Emily made a bunch of stickers that read "I Closed Brewer's Table," but they don't have the same jovial pride to them. Other than the stack I'll take home in my server's book, I'll never see another one again.

In the beer garden down below, Pleezer, a Weezer cover band that was here earlier this summer, is back, cranking the crowd up with the chunky power chords from "Buddy Holly."

"The other way I think about it," I say, "is that this was kind of like Woodstock. You know, it was this unprecedented party that, like all great parties, couldn't last forever—it had to come to an end for it to have been a party in the first place—not because it was a sucky party but, you know, because it was so great and intense and . . ."

Jorge kind of glances at me, and I'm arrested—I had forgotten how intense his eyes are. They're this strangely intense gray-blue that re-mind me of the way a cloudless sky in winter can be more profound and evocative than any simple blue one in summer.

"Anyway," I say, trying to wrap up, "it's like it was just this brief thing, but everybody who went there—everybody who was a part of it—none of them, none of us will ever be the same again."

The live music today is not for us, of course, but for the Minnesota United soccer fans who have begun streaming in with their stupid sum-mer scarves in advance of tonight's game. What I don't say to Jorge is that this crowd below, much like the one at any given huge music festi-val, is probably exactly what has kept most people from dining with us. They make parking a nightmare. Everybody needs to be ID'd and get their hand stamped regardless of whether they are college kids look-ing to get hammered or veteran fine diners eager to try out our restau-rant. Many regard it as a part of the experience, but just as many, if not more, find it insulting. On a night they are planning to attend the symphony, at dinner with us beforehand they have to walk a mile from where they were finally able to park, only to have to endure long lines to get through security at the front door, and then, wading through thick throngs of beer-drunk people, they endeavor just to find their way to the stairs that finally lead to our restaurant. It's as though they've got to survive a night at Burning Man or Coachella just to get to the really nice fine dining restaurant that's supposed to be hidden just beyond the main stage. Not everybody is up to that kind of hassle, yet were it not for the throngs in the beer hall and garden, we would never have had the financial largesse to do what we did upstairs in the first place. Chicken. Egg. Only in this case, the chicken is killing its baby chick in order to turn the nest into a strip club—according to Sadie's theory—called Surly's Gurlies or—according to Jorge's best guess—just another pizza place.

"Anyway," Jorge says. "It's been pretty cool."

The patio door opens and Cash comes out to join us. His eyes too are raw and he has this countenance that seems to say that he feels like he's missing out on whatever Jorge and I have been talking about.

"What's up, brother?" I say.

Cash, much in the way that I began with Jorge, can't say anything, but his face contorts itself in an effort to not again break down into tears.

"Yeah, fuck," Jorge says, acquiescing, almost, to my and Cash's weepiness. "I'm trying to wait until I'm good and fucking drunk before I start crying—again, I mean. Probably not healthy."

Instead of all crying together, we laugh, and for the moment, I think I'm going to make it through this night.

"Well," I say, gesturing to my ridiculous shorts-and-shirt-and-clogs attire, "I reckon it's time for this Dutch Boy to go get proper for work." And, with that, Pleezer begins playing "Say It Ain't So" and in we all go.

The evening will eventually begin, and the first of my last tables will be almost exactly the same as the first table I served on the first day we were open. An elderly couple who doesn't want to be bothered with who we are or what we have to say as servers. The way they pick up their menus and immediately turn them over, disappointed obviously with nothing more on the back—it tells the whole story of not just how the rest of the night will go but how our tenure at The Brewer's Table was doomed from the beginning. We gave it our absolute all, and to some it was epiphanic—food and drink at the level of transubstantiation. To most, however, it wasn't enough. "This is all you have?" my first/last couple demands. "And you don't have wine?"

Ordinarily I'd try to nuance my way through this now very old and familiar question. Tonight, "no" is all I say.

Charles comes over just then to drop off the complimentary amuse-bouche. They had been increasingly delicate and beautiful and had become de rigueur for anybody ordering the tasting menu. My favorite recent one was a few communion wafer–thin, house-made potato chips with crème fraîche and, as a throwback to our opening days, herring and flying fish roe. Today, Charles drops off a little bowl with three little golden orange orbs in it.

"May I present," he says, mock-formal, "cheesy poofs," and away he goes, leaving me to address my guests' baffled looks. They are, of course, straight out of the big plastic tub of Utz Cheese Balls that Jorge has had stocked, visible on the line to everyone, from the very beginning to the very end.

"Bon appétit!" I say and walk away.

The rest of the shift will be an absolute blur.

Before six, our entire space will be filled with people—people not just sitting at tables but people three-deep at the bar, people so dense along the window overlooking the brewery that nobody can get through them to serve anything—people everywhere all night long. Many friends and family, but mostly strangers who have the dangerous and toothy look of people about to loot the place.

Throughout the restaurant, a bunch of us have hidden Smirnoff Ices—these god-awful malt drinks that remind me of that abomination from the nineties, Zima. The recent tradition had been to "Ice" somebody on his or her last day, and the way it goes is that if that person finds one of the hidden bottles, they have to drop to one knee and down the whole terrible room-temperature thing in one swallow. It was understood that this was illegal and strictly against actual Surly policies and probably restaurant law, but this tradition was for somebody with one foot already out the door. I think it began with Miko's last day and then when Tiny Dancer left to go to HVAC school, and after it was announced we were closing, it happened at least once a week. Tonight was, of course, everybody's last day, and so what the hell did it matter?

There were two empty cases of Ice in the boardroom, and so the bottles lay in wait, barely hidden, throughout the restaurant for nearly everyone.

Right after the restaurant completely fills, I open our little reach-in cabinet for the caddies with the Sweet-and-Low that my elderly couple had just demanded only to find the peach varietal of Ice staring me dead in the face. I would've been happier to meet a cobra there, but the rules are the rules, and I take a knee and slam the thing as Cash and Emily clap and yell "Ice, ice, baby!"

It doesn't help my stomach any, but it does settle my nerves. I find another one in the silverware cabinet only five minutes later, but I glance around and nobody's seen me, so I shut the drawer and run over to Charles on the line.

"If you have a second," I say, knowing that Charles will make time and space expand for whatever anybody ever needs, "would you grab a spoon for seat 1 at 62? I've got to grab their coffee."

"Sure!" he says and zooms over to the cabinet. I don't move and instead just watch from the line, and Dustin and Jorge seem to have known what was going on, and we all cheer as Charles opens the drawer, and he turns around, initially shocked and bothered like he's been caught in a trap, but then he smiles and drops to a knee, and down the Ice goes.

Somehow the rest of the night happens and nobody gets hurt, and before we know it the last guests have been served and we're officially closed, but it wasn't like in a New Year's Eve countdown kind of way and so the restaurant is still full of guests and friends and beer hall people and us Brewer's Table people. Dan has ordered, of all freaking things, Domino's, and soon a stack of pizzas three feet high is at the pass where they have replaced Jorge, who has, apparently, already gone, and we're now all drinking as we eat the worst food that's ever been served in our restaurant. Some of us are still endeavoring to half-heartedly clean up, but it's a losing battle as the kitchen has decided en masse that they're done and aprons and kitchen towels are being whirled around and "Scat Man" is blaring from the sound system for the very last time.

Many, I'm sure, will stay until three or four in the morning. I won't be one of them, and I decide to quietly grab my things without a word to anybody and dash out the back door as though I'm just going to the bathroom but instead ghost off into the night on my bike.

There will be a party tomorrow hosted by Sadie and her husband, Dane—half owner of another Beard-nominated restaurant in Minneapolis. We'll almost all show up and we'll almost all be so hungover or blue or both that Dane calls his business partner to urgently bring an industrial blender from their restaurant and whip up batch after batch of margaritas and to force them into everybody's hands to prevent our collective sadness from creating a Minneapolis-sized black hole from sucking us all in. To a small degree, it almost works.

It'll be a potluck party and Dan will bring his special Paloma—a grapefruit and tequila mix that is a perfect way to follow up the frozen drinks we all know would soon lead to no good end—and Dane has smoked brisket and ribs and I'll bring a version of my favorite Paul Prudhomme recipe, Hoppin' John. Like putting his name on his shirts and underwear before we send our son to summer camp, Jenae had put my name on the ladle and the pot—not so everybody would know I made it but so we'd get the equipment back. Still it comes as a shock when, from behind, Jorge taps me on the shoulder.

"You make the Hoppin' John?" he says, an empty bowl in his hand.

I'm too sad and exhausted and not nearly drunk enough to make up a good lie, so I just shrug and say yeah.

"Pretty good," he says and then turns back to Tess and Dustin, where they remain, leaning coolly against the wall in the light rain, in my memory, like a classic and yet still modern frieze of everything and everybody I love at Surly.

There will be some weird Scandinavian yard game called Kubb that involved a lot of dangerous flying wooden objects, and there will be time with Rhonda, Sadie and Dane's horse-sized mastiff puppy, and there will even be karaoke where Emory, with the help of our other host, Toni, will struggle through most of the surprisingly complex "One Way or Another" by Blondie. But then Rhonda will momentarily get loose in the alley behind their house, and as the tender heart he is—no matter that the pup will come back safe and sound shortly thereafter—for Emory this is too much.

Too much because everything has been too much these last few hours, days, and months, and it will be impossible for somebody as smart and receptive as he is not to begin, finally, to suffer from it too, Stockholm Syndrome style, and so, too early, but really too late, we must go.

There will yet be the other restaurant where I'll get to work again with Cash, Celia, and Levi in the beautiful, art deco space where La Belle Vie was for years, but it too—like the lakeside place, like The Brewer's Table, like most restaurants—will close without much fanfare at all. It's difficult to be there for it—serving these last suppers, putting the lock on the door of all of our restaurant dreams—and it's difficult to think it doesn't have something to do with me, but that's pure egotism. Conventional wisdom says 90 percent of restaurants close within their first year. By that measure I guess we've succeeded at something, but despite the fact that we are trying today to have party to celebrate what we've done at Surly, to me it can't help but feel more like a wake.

And as we drive away from their home in Minneapolis, Jenae riding in the back of our car to try to console our boy, it takes nothing to remember riding home the night before, the mood and the darkness bleeding still together.

I never liked The Doors, having been born, I suppose, just a little too late to ever be fully under their sway. But it was impossible for me to not hear Morrison's chant in time with the cadence of my pedals last night—with the out-of-round bump of our tires tonight—both like the *whomp, whomp, whomp* of the helicopter blades at the end—or was it the beginning?—of *Apocalypse Now*. "This is the end, beautiful friend, this is the end," and down the road, toward home, alone, I go.

Acknowledgments

This book would simply not be a book without the support and encouragement of my friends, my former students, my current teachers, my writers' guild: Amy Vander Heiden, Josh Johnson, Kendra Tillberry, Pierre MacGillis, and Michael Daugherty.

And for inspiration and guidance from my fellow authors, I am forever indebted to Peter Geye, Adrienne Brodeur, Benjamin Percy, Nathan Hill, Michelle Wildgen, Michael Ruhlman, Brad Zellar, Jacob Paul, Nicole Walker, David McGlynn, Margot Singer, Stephen Tuttle, and Bruce Machart.

And to the dream team at the University of Minnesota Press who have been an absolute joy to work with from beginning to end, my deepest gratitude goes to Erik Anderson, Emily Hamilton, Laura Westlund, Shelby Connelly, Maggie Sattler, Daniel Ochsner, Anne Taylor, and Heather Skinner.

Without the crew at the Surly Brewing Company and the Brewer's Table, likewise, this book would be a ream of blank paper. To all the dishwashers, cooks, bartenders, my fellow servers and especially Dan DiNovis, Jorge Guzman, Emily Garber, Dustin Thompson, Linda and Todd Haug, and Omar Ansari, I owe you some of the best years of my life.

I am also indebted to my colleagues at the University of St. Thomas and the Center for Faculty Development there, as well as the generous support of the National Endowment for the Arts, the McKnight Foundation, the Aspen Writers' Foundation, and Isa Catto and Daniel Shaw of the Catto Shaw Foundation.

Melanie Rae Thon, my mentor. Lance was right. You are the Poet Laureate of the heart.

And Jenae and Emory, you're the names of everything I love.

Restaurants are ephemeral places. What they serve, where they are, how long they'll be around, who works in them can all change without notice. Some assert—like that Red Lobster in Duluth's Canal Park—with a brass plate bolted to the brick front of the restaurant that the name of the chef or manager is never going to change. Others insist on their menu or their website who their founder or their executive chef is. But the people who are the lifeblood of restaurants rarely get more than even a mention to the general public. If they do, it's usually just their first name and maybe a jaunty last initial. Accordingly, servers, busboys and busgirls, dishwashers, and prep cooks enjoy a kind of anonymity that, I think anyway, is both a kind of shame but also a kind of shelter. As in you can work in a restaurant as service staff and still maintain, if you so choose, another, separate career. At any rate, the names of the management teams in this book are completely real but the names of all servers, bartenders, bussers, cooks and dishwashers have been changed to help protect their identities. But I will never forget who you all are.

Matthew Batt is author of the memoir *Sugarhouse,* and his fiction and nonfiction have been published in the *New York Times, Outside* magazine, the *Huffington Post,* and *Tin House.* The recipient of grants from the National Endowment for the Arts, the McKnight Foundation, and Aspen Words, he teaches creative writing and lives with his family in Saint Paul, Minnesota.